Textbooks in Telecommunication Engineering

Era of the Internet and Network Science/Engineering

Series Editor

Tarek S. El-Bawab, Boston, MA, USA

Telecommunication and networks have evolved to embrace all aspects of our everyday life, including smart cities and infrastructures, healthcare, banking and businesses, manufacturing, space and aviation, meteorology and climate change, oceans and marine life, Internet of Things, defense, homeland security, education, research, social media, entertainment, and many others. Network applications and services continue to expand, virtually without limits. Therefore, specialized telecommunication and network engineering programs are recognized as a necessity to accelerate the pace of advancement in this field, and to prepare a new generation of engineers for imminent needs in our modern life. These programs need curricula, courses, labs, and textbooks of their own.

The IEEE Communications Society's Telecommunication Engineering Education (TEE) movement, led by Tarek S. El-Bawab- the editor of this Series, resulted in recognition of this field of engineering by the Accreditation Board for Engineering and Technology (ABET) on November 1, 2014. This Springer Series was launched to capitalizes on this milestone. The Series goal is to produce high-quality textbooks to fulfill the education needs of telecommunication and network engineering, and to support the development of specialized undergraduate and graduate curricula in this regard. The Series also supports research in this field and helps prepare its scholars for global challenges that lay ahead. The Series have published innovative textbooks in areas of network science and engineering where textbooks have been rare. It is producing high-quality volumes featuring innovative presentation media, interactive content, and online resources for students and professors.

Book proposals are solicited in all topics of telecommunication and network engineering including, but not limited to: network architecture and protocols; traffic engineering; network design, dimensioning, modeling, measurements, and analytics; network management and softwarization; cybersecurity; synchronization and control; applications of artificial intelligence in telecommunications and networks; applications of data science in telecommunications and networks; network availability, reliability, protection, recovery and restoration; wireless communication systems; cellular technologies and networks (through 5G, 6G, and beyond); satellite and space communications and networks; optical communications and networks; heterogeneous networks; broadband access and free-space optical communications; MSO/cable networks; storage networks; optical interconnects; and data centers; social networks; transmission media and systems; switching and routing (from legacy to today's paradigms); network applications and services; telecom economics and business; telecom regulation and policies; standards and standardization; and laboratories.

Proposals of interest shall be for textbooks that can be used to develop university courses, either in full or in part. They should include recent advances in the field while capturing whatever fundamentals that are necessary for students to understand the bases of the topic and appreciate its evolution trends. Books in this series will provide high-quality illustrations, examples, end-of-chapters' problems/exercises and case studies.

This series is indexed in Scopus. For further information and to submit proposals, please contact the Series Editor, Dr. Tarek S. El-Bawab, telbawab@ieee.org; or Mary James, Executive Editor at Springer, mary.james@springer.com.

Oscar Mauricio Caicedo Rendón

Machine Learning for Network Management

Springer

Oscar Mauricio Caicedo Rendón
University of Cauca
Popayán, Colombia

ISSN 2524-4345 ISSN 2524-4353 (electronic)
Textbooks in Telecommunication Engineering
ISBN 978-3-032-15175-9 ISBN 978-3-032-15176-6 (eBook)
https://doi.org/10.1007/978-3-032-15176-6

This Springer imprint is published by the registered company Springer Nature Switzerland AG
The registered company address is: Gewerbestrasse 11, 6330 Cham, Switzerland

I dedicate the book "Machine Learning for Network Management" to Lore, Pama, Samu, Edgar, Nina, Javier, Pilar y Julián. Without them, this book would never have been born.

"Los seres humanos no nacen para siempre el día en que sus madres los alumbran, sino que la vida los obliga a parirse a sí mismos una y otra vez".
Gabriel Garcia Marquez—GABO

Note from the Series Editor

This Series has its roots in the Telecommunication Engineering Education (TEE) movement in the USA, which I had the honor of leading from 2008 to 2015. The movement resulted from the remarkable impacts telecom networks had on society and the many more impacts looming on the horizon at that time. Major changes were unfolding in the theoretical foundations of this field of engineering, its design strategies, and practical applications. The industries we knew as the Telecom, Cable, and Data-Communication industries through the twentieth century were undergoing innovative technical changes, repositioning themselves in the market, and converging; in preparation for the global economy of the twenty-first century.

The TEE movement was prompted by a personal experience I went through when moving from industry to academia. Telecom engineering, or network engineering—since the network has been central to the evolution telecom has undergone for three to four decades- became a distinct multidisciplinary field of study. This field blends ingredients from electrical engineering, computer engineering, computer science, and other disciplines, but it should not be taught merely as a branch of one of them. It became evident that society is better served by adding specialized undergraduate/graduate courses and programs in telecom/network engineering to our higher-education system, especially in the USA where this is uncommon.

The story of the TEE movement has been told in the literature (e.g., IEEE Communications Magazine, Vol. 53, No. 11, pp. 35–39, November 2015). The movement accomplished its main goal of adding telecom/network engineering to the ABET list of recognized engineering disciplines. ABET is the Accreditation Board of Engineering and Technology, Inc., which accredits engineering programs in the USA and more than 40 other countries. However, a lot of work had to be done to capitalize on that milestone. The literature of this discipline was rich in scholarly research papers, but not as rich in textbooks designed to support the development of new university courses and curricula. These textbooks are needed to facilitate rigorous academic training for a new generation of network engineers. The idea of this Series arose subsequently to help the telecom/networking community overcome this shortcoming. Since its inception, Springer Nature's TTE (Textbooks in Telecommunication Engineering) has been doing this job. The proximity of the acronyms herein—between TEE and TTE—is coincidental, but worth noting!

I am proud of the inroads TTE has made in publishing quality textbooks and in fulfilling the goals and objectives the Series has had since its launch. By mid-2026, we approach the milestone of having nearly 50 books, including 30 published and about 20 in the making (whether in the writing, peer-reviewing, or production phases). More than 65 authors have contributed to this Series so far and they come from nearly 25 different countries, cultures, and higher-education systems, scattered all over the world.

The topics covered by this Series so far span the entire landscape of telecom and network engineering, including wireless, optical, satellite; access, metro, and core; terrestrial, space, aerial, and under-water; devices, systems, and networks; signal processing; queuing theory, traffic engineering, and operations research; network management; Internet and Internet of Things (IoT); cybersecurity; applications of data science, artificial intelligence and machine learning in network engineering; applied electromagnetics and antennas; and economy/business aspects of telecom. The Series includes a communication-systems primer, a radio-receivers primer, and a cellular-generations primer, for undergraduate and graduate students. A synchronization primer, Wi-Fi primer, and a cybersecurity primer are close to finishing, and others are expected.

We have covered cutting-edge topics with modern textbooks designed to propose innovative academic courses and programs, bearing in mind sometimes that further breakthroughs are imminent in these topics. Many of our books are also ideal for short courses and training of practitioners and industry executives to catch up with the rapid progress in the field. These include books discussing the roles of edge and cloud computing; blockchains; and AI/ML applications in telecom and networks. Special topics such as railway telecom, navigation systems, semantic communications, low-altitude wireless and video analytics are covered. Other topics at the forefront of contemporary socioeconomic challenges, such as broadband access, the digital divide, industry 4.0, and applications of IoT in various aspects of modern life are addressed.

I have tried my best to make this Series a top global de-facto home for quality telecom/network engineering textbooks, and I hope that many in our community consider it as such. The fruits of the hard work put in by many people to achieve this goal are now ready for our community to reap.

I want to thank my wife Shahira and my daughter Nadine for their support of my endeavor with this Series over more than a decade so far. I have put a lot of time into this mission and they have understood how important this work is to me. I owe a great deal to all the authors of this Series, without whom we could not build up such a monumental body of knowledge. Words cannot describe my appreciation of their contributions and talent. I would like to also express my thanks to Springer Nature, and to Mary James in particular. Mary has been my main Springer partner throughout this journey, and I couldn't have thought of a better one. Finally, I thank all my colleagues in the field of telecom/network engineering and our readership for their continued support of this global project.

Tarek S. El-Bawab, Ph.D., FIEEE
Series Editor

Preface

Why to Read This Book?

Although there are many scientific papers on using machine learning for network management, no book has compiled the landscape of machine learning to manage networks at a single point. In a nutshell, you need to read this book because:

- It showcases the intersection between machine learning and network management paradigms.
- It includes methodologies and guidelines practical to realize machine learning-driven network management.
- It provides a landscape of the machine learning techniques more relevant to network management.
- It showcases machine learning-based network management solutions.

What's Inside Machine Learning for Network Management?

Welcome to the book: *Machine Learning for Network Management*. This book has been lovely and meticulously prepared to engage readers eager for knowledge related to use machine learning techniques in the network management domain.

• Part I: Preliminaries

- Chapter 1. Introduction.
- Chapter 2. Network Management—A Primer.
- Chapter 3. Machine Learning in Networking—An Overview.

• Part II: Network Management Assisted by Learning Models

- Chapter 4. Network Management Assisted by Supervised Learning.
- Chapter 5. Network Management Driven by Unsupervised Learning.
- Chapter 6. Deep Reinforcement Learning for Network Management.
- Chapter 7. Network Management Assisted by Generative Learning.

• Part III: Exemplifying Machine Learning in Network Management

- Chapter 8. Improving and Explaining Learning Models

How to Read This Book?

This book's first part comprises the Chaps. 1–3. These three chapters present a self-contained overview of network management and machine learning domains, including concepts, terminology, and methodology. These chapters set the stage for the book's second and third parts, including Chaps. 4–8. After completing the first part, instructors are recommended to cover Chaps. 4–8 sequentially; each of these five chapters leverages material from the preceding ones.

Audience

This textbook is for an elective course that addresses network management with machine learning techniques. It can be used in computer science and electrical engineering departments. Regarding programming languages, the book assumes that the student has experience with Python. The book is appropriate for undergraduate and first-year graduate courses. It should also be helpful to practitioners in the networking industry.

Supplements for Instructors

The book offers a complete supplementary package (including source code of examples) to aid instructors in teaching machine learning in the network management domain. This material can be accessed from the Instructor Resource Center (http://www.mlfnm.com). Visit the Instructor Resource Center for information about accessing these instructor's supplements. PowerPoint slides are also provided for all six chapters. The slides cover each chapter in detail. They use graphics and animations (rather than relying only on monotonous text bullets) to make the slides interesting and visually appealing. The original PowerPoint slides are provided so you can customize them to suit your teaching needs best.

Popayán, Colombia

Oscar Mauricio Caicedo Rendón

Acknowledgments A BIG THANKS to everyone who has helped me from the earliest drafts of this book up to this final edition. Also, I am very thankful to my research colleagues and under and post-graduate students who have helped with their investigations to create the knowledge contained in this book. Special thanks go out to:

Nelson Fonseca, Stadual University of Campinas
Raouf Boutaba, University of Waterloo
Lisandro Granville, Federal University of Rio Grande do Sul
Telematics Department, Universidad del Cauca
COMSOCAUCA, Universidad del Cauca

Competing Interests The author has no competing interests to declare that are relevant to the content of this manuscript.

Contents

About the Author

Oscar Mauricio Caicedo Rendón is a full professor at the Universidad del Cauca, Colombia, where he is a member of the Telematics Engineering Group of the Telematics Department. He received his Ph.D. degree in computer science (2015) from the Federal University of Rio Grande do Sul, Brazil, and his M.Sc. in telematics engineering (2006) and his degree in electronics and telecommunications engineering (2001) from the Universidad del Cauca. He serves as Associate Editor for the *IEEE Communications Magazine Series on Network Softwarization and Management*, the *IEEE Networking Letters*, the *IEEE Latin America Transactions*, the *IEEE Transactions on Cognitive Communications and Networking*, the *IEEE Transactions on Network Science and Engineering*, and the *IEEE IoT Journal*. He served and keeps serving as Technical Program Co-chair for IEEE-sponsored international workshops and conferences, including the Machine Learning for Communications Symposium of the IEEE Global Communications Conference and the IEEE Latin-American Conference on Communications. He has also been Technical Program Committee member and reviewer of well-known conferences and journals, such as NetSoft, EuCNC, ISCC, COLCOM, NOMS, CNSM, TNSM, JONS, COMNET, TNSE, TCCN, and ESWA. His recent research interests include network and service management, network functions virtualization, software-defined networking, artificial intelligence and machine learning for network management, and network softwarization. e-mail: omcaicedo@unicauca.edu.co

Acronyms

ARF	Adaptive Random Forest
CNN	Convolutional Neural Network
CRISP-DM	Cross-Industry Standard Process for Data Mining
CRISP-ML(Q)	CRISP ML with Quality
DDQN	Double Deep Q-Network
DNN	Deep Neural Network
DQN	Deep Q-Network
FNN	Feed-Forward Neural Network
GAN	Generative Adversarial Networks
GNN	Graph Neural Network
KDD	Knowledge Discovery in Database
KPI	Key Performance Indicators
LIME	Local Interpretable Model-agnostic Explanations
LSTM	Long Short-Term Memory
MDP	Markov Decision Process
MUSE	Model Understanding through Subspace Explanations
PPO	Proximal Policy Optimization
RBM	Restricted Boltzmann Machines
RF	Random Forest
RNN	Recurrent Neural Network
S3VM	Semi-Supervised Support Vector Machine
SARSA	State–Action–Reward–State–Action
SHAP	Shapley Additive exPlanations
SVM	Support Vector Machine
VAE	Variational AutoEncoder
XAI	eXplainable Artificial Intelligence
XDRL	eXplainable DRL
XRL	eXplainable RL

Part I
Preliminaries

The book's first part presents the main concepts related to network management and machine learning and their conjunction. In this way, Chap. 1 introduces this book, comprising motivation and organization. Chapter 2 reviews overall network management and its variations, including functional areas, cognitive network management, and zero-touch network management. Chapter 3 overviews machine learning paradigms, algorithms, and methodologies.

Introduction 1

1.1 Motivation

Given the growing complexity of current networks and the increasing importance of connectivity for individuals, network management tasks have become pivotal. To highlight the relevance of network management, let's consider: a hospital in which the electronic health records of patients must be retrieved and analyzed instantly during emergencies, a university campus where thousands of students rely on stable connectivity to attend virtual lectures and run generative artificial intelligence queries, businesses that depend on secure online transactions, and millions of people expecting uninterrupted video streaming when massive events happen. The success of these scenarios depends on a complex network of routers, switches, servers, and protocols that must operate seamlessly. However, networks are not static; in fact, they are highly dynamic and need to face constant change and stress, including sudden traffic surges during live events, cyberattacks probing for vulnerabilities, software updates that risk service interruptions, and the ongoing demand for higher speeds at lower costs. Without effective management, these challenges result in dropped connections, data breaches, financial losses, and even risks to human life in critical applications such as remote surgeries.

From a high-level abstraction, network management can be understood as a field that ensures networks are reliable, secure, adaptable, and efficient. Remarkably, nowadays—featured by cloud computing, the Internet of Things, and 5G/6G mobile technologies—network management goes beyond a background field; it is a strategic enabler of innovation and a key factor in building trust among businesses. This book guides the reader through the concepts, methodologies, and emerging approaches that enable performing network management assisted by machine learning. We will explore not only the "what" of technologies and protocols but also the "why"—the reasons they matter and the role they play in building modern communication networks. In this regard, the following subsections of this introductory chapter provide a brief overview of network management and its relevance, machine learning (ML) and its importance, and the intersection of network management with ML.

1.1.1 Network Management

From a more formal perspective, the network management field encompasses monitoring and controlling, planning extensions or modifications, and incorporating new elements into an existing network without disrupting ongoing services running on top of it. In traditional networking, network management was performed on a per-service basis, as a specialized network was built for each service. With the convergence and continuous growth of networks, as well as the emergence of new services, overall network management has become increasingly complex daily.

From a classical perspective, network management approaches use remote invocation or delegation. Network management based on remote invocation typically follows a manager/agent architecture, in which the manager acts on managed objects through an agent. In particular, the manager collects information, issues commands, and provides operators with a global view of the network, and the agent maintains local information and responds to requests from the manager. One of the most relevant network management solutions is the Simple Network Management Protocol (SNMP); remarkably, SNMP-based solutions follow the manager/agent architecture. The Network Configuration protocol (NETCONF) is another network management solution that exemplifies the use of the manager/agent architecture. SNMP primarily focuses on monitoring, and its configuration capabilities are more limited compared to modern protocols. NETCONF was designed to overcome

Supplementary Information The online version contains supplementary material available at https://doi.org/10.1007/978-3-032-15176-6_1.

O. Caicedo Rendon, *Machine Learning for Network Management*, Textbooks in Telecommunication Engineering, https://doi.org/10.1007/978-3-032-15176-6_1

the limitations of older protocols, such as SNMP; it uses data models based on eXtensible Markup Language (XML) to describe device configuration and YANG as a data language to model state information. A variation of the remote invocation architecture occurs when invocations on managed objects are carried out through interfaces or objects provided by distributed technologies, such as the Java Remote Method Invocation (JRMI) and the Common Object Request Broker Architecture (CORBA). JRMI is a mechanism for building distributed applications in the Java programming language, where an object running in one Java Virtual Machine (JVM) invokes methods on an object running in another one, even if the JVMs are operating on different physical machines. CORBA enables communication of distributed object-oriented applications independently of their implementation language; JRMI is Java-tight.

In network management by delegation, emerged as networks became more complex and centralized control of manager/agent architecture became a bottleneck, there is code mobility that implies the management intelligence or task logic (e.g., fault detection and configuration adjustments) travels from the central manager to the agents or intermediate entities; instead of waiting for instructions, agents (or local managers) can autonomously handle specific tasks. The trip can involve just one hop, as in the manager/agent architecture, or multiple ones, as in mobile agents, where the logic travels between agents following a predefined or autonomous itinerary. In network management based on mobile agents, agents perform tasks and operations across different network parts, often to enhance flexibility (context awareness allows for fast adaptation to changing network conditions), reduce latency (carrying out tasks closer to the data source decreases the time spent for decision-making), and minimize the communication overhead between devices (local processing data can be helpful to reduce the amount of data transmitted to central management systems).

With the emergence and evolution of Internet technologies, network management architectures based on Web Services have emerged to complement the capabilities offered by delegation and remote invocation-centered management solutions. Those architectures comprise three layers. The upper one uses web browsers as a standard graphical user interface, often based on HyperText Markup Language (HTML), XML, Cascading Style Sheets (CSS), JavaScript, and JavaScript Object Notation (JSON), allowing network practitioners to configure and monitor network devices, for instance. This layer interacts via the Hypertext Transfer Protocol (HTTP) with a Web Service Requester (acting as a centralized manager that offers network management logic) located in the middle layer, jointly with a set of Web Service Providers (acting as agents). The Requester and Agents communicate through HTTP. In particular, HTTP requests and responses contain XML data for Web Services based on the Simple Object Access Protocol (SOAP) or JSON for Web Services that operate according to the Representational State Transfer (REST) architectural style. The agents in the middle layer represent the network management agents that operate in the bottom layer, utilizing protocols such as SNMP, NETCONF, RESTCONF, and gNMI (Google Remote Procedure Call Network Management Interface).

From a practice perspective, network management applications involve multiple services. In this regard, Web-based network management applications are convenient since they can combine various web services to generate a new composite network management service. Choreography and orchestration are valuable concepts for achieving this composition. Choreography describes the collaboration between the Web Services participating in the composition without specifying their internal behavior; the Web Service Choreography Description Language (WS-CDL) aims to coordinate interactions among Web services and their stakeholders in a decentralized manner. In Web Services orchestration, the participating services are controlled by a single endpoint central process (another Web Service); the Business Process Execution Language (BPEL) allows for defining complex tasks as a combination of Web services. Composite network management applications can be achieved statically through scripts developed at a low level by network practitioners or with mashup tools at a high abstraction level, and dynamically through Artificial Intelligence (AI) technologies. In the composition based on static scripts, the network management logic is written directly in a programming language such as Java or Python. A mashup tool allows developing mashups that are composite Web applications created by combining different resources available on the Web and centered on end-users; end-user-centric means that mashup-based network applications may be produced by users who usually have non-advanced programming skills, and further, mashups encourage cooperation among end-users and the reuse of existing Web-based network management applications.

Network Management—Relevance

The role of network management extends beyond technical operations to have a direct impact on productivity, customer satisfaction, and overall business success. Notably, dynamic, intelligent, and autonomous network management is critical for ensuring uninterrupted access to network resources and services. It is also pivotal for ① monitoring and adjusting traffic flows to prevent bottlenecks and improve performance; ② identifying and addressing potential issues before they turn into significant outages, detecting suspicious activities, malware, or intrusions in real-time; ③ handling user and device permissions to prevent unauthorized access; ④ guaranteeing the meeting of regulatory standards to avoid legal and financial penalties; ⑤ Dynamically allocating computational and networking resources to accommodate growing demands; numrounded6 facilitating the seamless addition of new devices, applications, and technologies while maintaining operational stability; ⑦ minimizing productivity losses and revenue impact caused by network outages; ⑧ offering informed decision-making for resource planning and upgrading, ensuring rapid restoration of services when failures or cyberattacks happen; ⑨ providing real-time insights about network health to prevent disruptions, reducing human involvement to automate repetitive tasks and avoid service disruption; and ⑩ offering seamless access to applications and services, improving productivity.

Remarkably, this book addresses intelligent self-driving networking by intersecting the fields of network management, traditional ML, Deep Learning (DL), and Generative Artificial Intelligence (GAI).

1.1.2 Machine Learning

ML is one of the most disruptive technologies of the modern era. In a nutshell, this field of AI enables the development of systems that learn and improve themselves from data and experience without being explicitly programmed. Inside ML, various learning paradigms have emerged, the most well-known of which are organized in chronological order of appearance: Supervised Learning, Unsupervised Learning, Reinforcement Learning (RL), Semi-supervised Learning, Deep Learning (DL), Deep Reinforcement Learning (DRL), Self-Supervised Learning, and Generative Learning. This book covers algorithms in all the above-mentioned paradigms, aiming to illustrate to the reader how to achieve intelligent autonomous network management.

Supervised Learning, whose first techniques were presented around the 1950s, is one of the earliest paradigms in ML; it uses datasets formed by a collection of samples embracing features (independent variables) and labeled data (target variable) to train models able to predict outputs, such as the type of network traffic or size of flow, for unseen inputs (samples). Unsupervised Learning, whose early algorithms were introduced in the 1960s, focuses on discovering patterns and structures, such as regular and abnormal traffic, in unlabeled data. RL, whose first algorithms were proposed in the 1970s, aims to learn to make decisions related to, for instance, optimal paths and resource allocation, by interacting with an environment that can represent a traditional or softwarized network. Semi-supervised Learning includes hybrid techniques that leverage labeled and unlabeled data to address the challenge of acquiring large amounts of labeled data—a typical problem in the networking area, for instance, when it is necessary to classify encrypted traffic. The first semi-supervised techniques appeared in the 1990s.

DL, a learning paradigm pivotal for image recognition, surged in the 2000s as a means of overcoming the limitations faced by traditional ML techniques when handling extensive sequential data, thanks to the resurgence of neural networks. In the network management area, DL allows handling the immense quantity of data generated by medium and large-scale networks. DRL, an ML approach that combines DL and RL techniques, emerged in the 2010s for solving robotics, gaming, and decision-making applications. In network management, DRL has been successfully employed to address routing and allocation problems in data center networks, for instance. Self-supervised Learning, whose first algorithms appeared in the 2010s, is a paradigm for pre-training large-scale models using unlabeled data with proxy tasks (e.g., predicting missing parts of data). Therefore, self-supervised techniques, which help understand complex traffic patterns and detect unusual activities in real-time, eliminate the need for labeled datasets. Notably, obtaining datasets in network management is costly and time-consuming. Generative Learning may be the most disruptive AI paradigm, enabling the creation of new data samples from an enormous given dataset; its first techniques were developed in the 2010s. Generative models are being explored mainly to achieve self-configuring networks. Remarkably, many ML paradigms are built upon previous ones, meaning that the most recent ML paradigms often leverage advancements in computational power, data availability, and theoretical understanding of earlier ones.

Machine Learning—Relevance

ML is the backbone of AI. In short, the relevance of ML is due to, among other aspects, it enables:

- Data-driven decisions refer to ML algorithms that can analyze massive datasets to uncover patterns and trends, and offer insights that humans might overlook. They also help organizations make highly accurate and actionable predictions (e.g., sales forecasts and customer behavior). Furthermore, they can process data quickly, enabling real-time decision-making in dynamic environments.
- Automation of repetitive and complex tasks. ML algorithms enable the development of applications for automating repetitive tasks, thereby freeing up human resources. They also allow automating tasks that require intelligence, such

as Natural Language Processing (NLP), image recognition, and speech translation. Additionally, they can help improve accuracy and reliability when performing complex tasks compared to manual processes.

- Customization relates to the power of ML algorithms in recommendation systems (when considering humans in the loop), which provide tailored content and product suggestions to individual users or sets of users sharing a profile. They also enable the improvement of marketing targeting by predicting customer preferences and behavior.
- Productivity refers to ML allowing the automation of mundane tasks, thereby permitting human workers to focus on creative and strategic activities. It can also identify inefficiencies in operations and suggest improvements; in this regard, it is remarkable that ML algorithms can handle growing datasets and complexity with minimal human intervention.
- Competitive advantage relates to Companies that can leverage ML to create innovative products and services faster than competitors; early adoption of ML positions businesses as industry leaders. In this regard, it is also remarkable that ML enables rapid adaptation to market changes and emerging trends.
- Driving innovation refers to leveraging the power of ML algorithms to advance outstanding technologies, such as 5G, 6G, Cloud and Edge Computing, and Augmented Reality. Furthermore, ML facilitates the creation of new business models and fosters innovation by enabling creative solutions to unique problems.

Beyond network management, some application domains (a.k.a. verticals) that have been leveraging ML techniques for a long time are (i) healthcare to improve diagnostics, drug discovery, and treatment planning, (ii) finance to detect fraud and assess credit risks, manufacturing to optimize supply chains and predict equipment failures, (iii) retail to enhance inventory management and improve customer engagement, (iv) transportation to offer power autonomous vehicles and optimize logistics, and (v) situational management, that was born in the military field, but that right now can be used to predict and mitigate the effects of disasters, such as earthquakes and hurricanes.
Overall and from a broad perspective, ML algorithms play a foundational role in accelerating scientific discoveries. In this regard, they enable the optimization of energy usage, reduction of operational costs, and minimization of environmental impact by automating processes and reducing errors, and they allow intelligent systems to understand and interact effectively with human language. From a narrow scope, in addition to the aspects already discussed, ML's importance lies in its current applications and potential to deal with the challenges of tomorrow's network management.

1.1.3 Machine Learning for Network Management

Relevance
Nearly, half of network issues are attributed to poor configuration or change management performed manually by network practitioners. In this regard, AI, especially ML, enables the creation of dynamically composite, intelligent, and autonomous network management applications pivotal for reducing human involvement in the networking domain. Dynamically composite means that the choreography or orchestration can be built in (near) real-time. Intelligence and autonomy refer that network management decisions are data-driven, which implies (i) collecting network status from many heterogeneous sources, (ii) correlating and fusing multi-source information to assist the recognition and comprehension of network status and, so support the timely and correct making-decision when network situations (e.g., a router has a sudden loss of received packages and a switch has an unforeseen loss of transmitted packages) happen, (iii) analyzing past network statuses to find patterns and predict future states, (iv) reasoning and planning based on descriptions, predictions, and prescriptions, and implementing actions to address network state within some predefined constraints; and (v) presenting information aiming at the human comprehension maximization about management decisions and their impact.
ML has become pivotal in enhancing networking since it enables various management approaches, such as autonomic, cognitive, and zero-touch network management. Autonomic network management refers to enabling networks that are capable of self-management, thereby reducing the need for human involvement. Self-management refers to the ability of networks to configure, heal, optimize, and protect themselves autonomously [1]. Cognitive network management involves perceiving current conditions, planning, deciding, and acting based on those conditions, and learning from the outcomes to improve future decisions, seeking to achieve self-aware and adaptive networks [2]. Zero-touch network and service management aims for fully automated network and service management without human intervention, which is crucial for handling the complexity of modern networks, especially with the advent of 5G, 6G, and beyond [3, 4]. Notably, in parallel with network evolution, network management have advanced toward increased automation and intelligence.

Challenges
Achieve ML-based dynamic, intelligent, and autonomous network management faces critical challenges. ① There is little data availability, so obtaining high-quality data for training, testing, and evaluating ML models for solving networking

problems is difficult. ② Labeling can be time-consuming and prone to errors because of the scarcity of labeled network data for supervised learning techniques. ③ The handling of sensitive network data while maintaining compliance with privacy regulations. ④ Data integration is highly complex as networks generate heterogeneous data from various sources (e.g., user activity and traffic logs). ⑤ The building up of cross-domain ML solutions is intricate since networks involve multiple network segments (e.g., Edge, Core, and Radio Access Networks), domains, vendors, and technologies; notably, the lack of standardized frameworks or protocols for ML-based network management hinders implementation and collaboration, leading to non-integration related to interoperability issues. ⑥ Large and dynamic networks need ML models to scale with increasing data volumes and changing network conditions; continuous updating and deployment of models is necessary under such conditions but without retraining from scratch, leading to streaming learning needs. ⑦ ML models trained for one type of network or use case may not generalize well to another, but the learning can be leveraged by transfer learning. ⑧ Performing network management tasks demanding real-time responses is challenging when coped with ML models with high computational requirements, specially in resource-constrained environments like IoT and Edge devices; in this regard the exploration of fine-tuned small models is a priority. ⑨ The balance between accuracy and fast response, ML modes with high accuracy may be computationally expensive and thus increase the energy footprint of network management, while faster models might sacrifice precision. ⑩ Complex deployment, although softwarized and virtualized networks facilitate deploying ML models into network management solutions, performing it into existing ones demands significant modifications. Besides, black-box models have a shortfall in interpretability, explainability, and transparency, making network operators hesitant to trust them. ⑪ As the development and deployment of ML-based network management solutions are expensive, they can be unfeasible for smaller organizations. In this path, it is also highlighted that few professionals are skilled in ML and network management, which can delay adoption and deployment of autonomous/self-driving networks.

1.2 Objectives

Unlike many books that cover the use of ML in diverse IT disciplines, this book focuses on putting on how ML is, can, and will be used in network management. In this regard, its primary goals are:

- Present network management aided by ML, including supervised, unsupervised, deep, generative, and by reinforce learning.
- Describe ML techniques in the context of the network management domain.
- Introduce methodologies to build up and explain ML-based network management.
- Showcase examples of ML-based network management solutions.

1.3 Organization

This book has been organized as follows.

Part I: Preliminaries

- **This chapter: Introduction.** This chapter presents the motivation on using ML to cope with network management problems, raises the associated pivotal challenges, and describes the overall structure of this book.

- **Chapter** 2. **Network Management—A Primer.** This chapter aims to provide an overview of network management paradigms. It presents the functional areas of network management from the perspective of the International Organization for Standardization, introduces cognitive network management, and discusses zero-touch network and service management.

- **Chapter** 3. **Machine Learning in Networking—An Overview.** This chapter intends to offer an overview of ML from the perspective of network management. It presents the ML paradigms and their potential for network management. Then, it reviews the evolution and key milestones of ML algorithms. Finally, it discusses methodologies for building and explaining ML solutions.

After this first part, each chapter introduces one or more ML algorithms and exemplifies their use from a practical perspective by including code snippets in the Python language. The snippets are available for download in the book's GitHub repository and can be executed in the Google Colab Platform. By modifying such codes, readers are encouraged to apply what they learn or explore further.

Part II: Network Management Assisted by Learning Models

- **Chapter** 4. **Network Management Assisted by Supervised Learning.** This chapter aims to provide the fundamentals for conceiving, developing, and testing tasks within network management functional areas using supervised ML techniques. In this direction, this chapter presents and exemplifies various algorithms, such as linear, Support Vector Machine (SVM), Decision Tree (DT), Random Forest (RF), eXtreme Gradient Boosting (XGBoost), Feedforward Neural Networks (FNN), Convolutional Neural Networks (CNN), and Recurrent Neural Networks (RNN), in the network management domain. This chapter also discusses (including implementation codes) using supervised learning to build diverse network management solutions: network latency prediction, network throughput prediction, anomaly detection, encrypted traffic classification, and bandwidth utilization.

- **Chapter** 5. **Network Management Driven by Unsupervised Learning.** This chapter intends to offer foundations to devise, develop, and evaluate tasks framed into network management functional areas with supervised ML techniques. In this regard, this chapter presents and exemplifies K-means, Hierarchical Clustering, Density-Based Spatial Clustering of Applications with Noise (DBSCAN), Principal Component Analysis (PCA), t-distributed Stochastic Neighbor Embedding (t-SNE), Isolation Forest, and One-Class SVM algorithms in tasks related to managing faults, configuration, accounting, performance, and security. This chapter also discusses (including implementation codes) using unsupervised learning to build diverse network management solutions: faults clustering, flows clustering, IoT device abnormal behavior detection, traffic anomaly detection in OpenFlow-enabled switches, and dimensionality dataset reduction for traffic analysis.

- **Chapter** 6. **Deep Reinforcement Learning for Network Management.** Managing complex network systems has become a formidable challenge in the rapidly evolving landscape of the networking domain. Traditional methods fall short in optimizing network management tasks in dynamic environments. This chapter introduces RL and DRL as complementary and revolutionary approaches to network management. It aims to provide knowledge to build up network management functional areas with RL and DRL. Initially, this chapter presents a taxonomy of RL and DRL algorithms. Then, it shows how RL and DRL can support various management tasks. Finally, it introduces a methodology to realize RL and DRL-based solutions; the use of the application is outlined by practical examples and their corresponding implementation codes.

- **Chapter** 7. **Network Management Assisted by Generative Learning.** As generative AI models are changing the world, this chapter aims to present them in the domain of network management. Initially, it discusses and exemplifies the use of Autoencoders and Variational Autoencoders to perform tasks within the functional areas of network management, specifically Faults, Configuration, Accounting, Performance, and Security (FCAPS). Then, it presents Generative Adversarial Networks (GANs) and illustrates their use for managing networks. Finally, it introduces the transformers' architectures (encoder-only, decoder-only, and encoder-decoder) and shows how they can be employed in the FCAPS.

Part III: Exemplifying Machine Learning in Network Management

- **Chapter** 8. **Improving and Explaining Learning Models.** This chapter aims to showcase the use of ML techniques in network management and explain how they can be explained and improved. Initially, it exemplifies the use of Random Forest, XGBoost, SVM, Autoencoders, CNN, and RNN for attack detection in IoT. Afterward, it shows how to enforce the SHapley Additive Explanations (SHAP) technique to show what features are used by the CNN model built for making decisions about attack detection. Finally, it presents the CNN improvement achieved by the SHAP analysis.

References

1. N. Samaan, A. Karmouch, Towards autonomic network management: an analysis of current and future research directions. IEEE Commun. Surv. Tutor. **11**(3), 22–36 (2009). https://doi.org/10.1109/SURV.2009.090303
2. M. Zorzi, A. Zanella, A. Testolin, M.F. De Grazia, M. Zorzi, Cobanets: a new paradigm for cognitive communications systems, in *Proceedings of the 2016 International Conference on Computing, Networking and Communications (ICNC)* (2016), pp. 1–7. https://doi.org/10.1109/ICCNC.2016.7440625
3. M. El Rajab, L. Yang, A. Shami, Zero-touch networks: towards next-generation network automation. Comput. Netw. **243**, 110294 (2024). https://doi.org/10.1016/j.comnet.2024.110294
4. O.G. Lira, O.M. Caicedo, N.L.S. da Fonseca, Large language models for zero touch network configuration management. IEEE Commun. Mag. **42**, 1–8 (2024). https://doi.org/10.1109/MCOM.001.2400368

2 Network Management: A Primer

2.1 Network Management Functional Areas

At the dawn of the 1980s, the International Organization for Standardization (ISO) presented the most well-known model for managing telecommunication networks and services. The model includes five functional areas (see Fig. 2.1): Faults, Configuration, Accounting, Performance, and Security. Each functional area copes with a particular aspect of network management to guarantee that networks and services operate efficiently, securely, and reliably.

2.1.1 Faults Management

This functional area of network management covers various phases to minimize errors, faults, fails, and failures in networks or network services and their respective effects on the availability and reliability of applications and services. A transient (e.g., an overwhelmed router) or persistent (e.g., a router table misconfigured) error refers to an incorrect or unexpected event that occurs within the network or one or more of its components; errors are recorded and monitored to avoid resulting in faults. A fault is a condition usually caused by an error or a series of errors that deviates a network component (hardware or software) from regular operation; a fault indicates something is wrong and can result in a fail if unattended. A fail (e.g., a congested link that eventually resumes its normal function without requiring a complete replacement) is a condition that prevents a network device or service from operating on time but does not result in a complete breakdown; thus, a fail may not require immediate action. A failure occurs when a network device (e.g., a router crash) or service stops (e.g., a Web server crash) functioning, necessitating an immediate intervention to return to normal operation; therefore, a failure is a more severe condition than a fail.

Functionalities

The functional area of fault management encompasses primarily the following functionalities.

- Detection. This functionality refers to detecting errors, faults, fails, and/or failures that may impact the performance of network devices or services. It demands a continuous network monitoring capability to capture data related to the above-mentioned problematic conditions and the network status; SNMP and OpenFlow are standardized protocols used to perform monitoring tasks in traditional and softwarized networks, respectively. The monitoring capability may involve active, passive, and proactive measurement tools. Active tools (e.g., ping and iperf) aid the fault management system in identifying and rectifying potential issues in real time, sometimes before they even become an error, fault, fail, or failure. The trade-off, however, is more network traffic. Passive monitoring tools listen for events (e.g., threshold violations) or analyze historical data from network logs to detect problems that have occurred. While reactive fault management, often supported by passive monitoring, is more conservative in terms of resource usage, it may detect faults too late. Proactive fault management (e.g., AI/ML-based fault detectors) can use collected data to foresee failures, such as predicting when a link turns on in a bottleneck.

Supplementary Information The online version contains supplementary material available at https://doi.org/10.1007/978-3-032-15176-6_2.

O. Caicedo Rendon, *Machine Learning for Network Management*, Textbooks in Telecommunication Engineering, https://doi.org/10.1007/978-3-032-15176-6_2

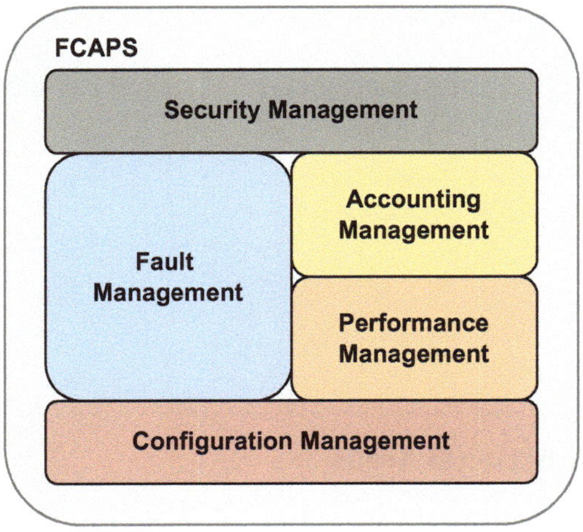

Fig. 2.1 Functional areas—network management

- Diagnosis. This functionality involves identifying the source of an issue (error, fault, fail, or failure) through a diagnostic testing capability and locating it within the network topology. Furthermore, diagnosis demands a comprehensive ability to classify the issues and correlate the corresponding causes; fault management systems often group related events for administrators and provide a root cause analysis.
- Isolation. This phase requires separating errors, faults, fails, or failures to prevent their spread and impact other devices, links, segments, or services operating in the network. In fault management, isolation can be manual, automatic, or autonomous. The manual one demands human skills, which can be prone to more problems. Automatic Isolation requires complex software-based systems built with programmatic rules. Autonomous Isolation claims the use of ML techniques and can be framed within concepts such as cognitive network management and zero-touch network management, which will be discussed later in this chapter. Those concepts aim to reduce human intervention during the execution of management tasks.
- Notification. This functionality refers to automatically informing network administrators via alerts and notifications about network activity, informing them about what is happening in the network. An alert is a real-time signal that prompts immediate action to ensure a timely response in both reactive and proactive fault management. A notification is typically a less urgent message than an alert (e.g., the email server is down); it can inform the receiver about the network status, routine events (e.g., scheduled maintenance), and changes (e.g., a backup link has been successfully established).
- Resolution. This functionality comprises performing manual (primarily when hardware problems happen) or software-based actions to repair (it does not mean the service is ready to use again) the problem's source or replace it, or implement and deploy a workaround.
- Restoration. With this functionality, once the problem is repaired, the incident is resolved, and lost data is recovered, the fault management system automatically runs predefined scripts or programs to get services up and running as soon as possible. Depending on the functionalities offered by the system, restoration may or may not require human intervention.
- Documentation. This functionality embraces the manual, automatic, and autonomous creation of reports describing the problems the network and its services suffered and the solution actions. Documentation is pivotal to improve fault management systems by learning from past.

2.1.2 Configuration Management

Configuration management ensures that the network operates efficiently and lays the foundation for all other functional areas of network management. A configuration management system can integrate manual, automated, and autonomous interventions to meet design requirements, QoS/QoE metrics, and SLAs. In manual configuration management, the network administrator performs highly complex and time-consuming tasks, which may lead to mis-configurations or insecure configurations, resulting in service disruptions or potential non-SLA compliance. An automated configuration management system is primarily software-assisted; therefore, network administrator intervention is reduced to minimize errors and rework. Autonomous configuration management utilizes AI/ML techniques to minimize or eliminate human intervention in configuration tasks, which will be discussed in the following subsections of this chapter from the perspectives of cognitive network management and zero-touch network and service management.

Functionalities

The configuration management area primarily involves the functionalities of inventory management, change handling, version control, auditing, and provisioning, all of which are related to tracking and controlling access to the network configuration based on roles.

- Inventory. This functionality refers to manually or automatically recording in detail all configurations (current and past) of all network elements (hardware or software). Inventory management involves recording who performed the initial network configurations and subsequent updates.
- Change handling. This functionality aims to minimize network disruptions by following a predefined process to realize controlled network changes. Such a process includes planning, testing, approving, and documenting any changes to the configuration of a device or service within the network. Handling management includes registering (documenting) who performed the network reconfigurations to maintain an updated inventory.
- Version control. This functionality involves maintaining a historical record of configurations to enable rolling back to a previous network state when necessary. The versioning control functionality is pivotal for decreasing service outage time.
- Auditing. The auditing functionality embraces the process of verifying that configurations comply with organizational policies, standards, and best practices. The configuration's verification encompasses both syntactic and semantic elements; Batfish and Plankton exemplify practical network configuration verification tools.
- Provisioning. This functionality refers to setting up devices and services with minimal or no human intervention using predefined templates, languages, policies, or configuration intents to ensure consistency and excellent network performance. In this regard, a configuration template can include specific runnable commands necessary to setup a router, a virtual private network or a firewall, for instance. A configuration language, such as JSON and YANG, can help to setup a network device or service by following a predefined syntax and semantics; these types of languages operate at a higher abstraction level than templates and need an agent for their execution (e.g., a NETCONF agent to run the configurations expressed by YANG). JSON is a lightweight, text-based data format used to structure, encode, and exchange information between network elements, management systems, and applications. YANG is a data modeling language (standardized in the RFCs 6020 and 7950) used to define the structure, constraints, and semantics of configuration and state data manipulated by network management protocols. Unlike JSON, YANG is about defining models that describe what data can exist, its hierarchy, and how it should be interpreted. A configuration policy provides a flexible, portable, and customizable solution for configuring one or more network entities on the fly. A policy can be defined using specific languages (e.g., Routing Policy Specification Language and Logic-based Policy Definition Language), which operate at a higher abstraction level than configuration languages. Policy configuration languages require specialized engines to translate the policies into actionable configurations. Intents are high-level, human-friendly policies or objectives that describe what the network should achieve, without specifying the exact means to achieve it. Policies and Intents will be discussed later in this chapter, within the context of cognitive network management and zero-touch network and service management, respectively.

2.1.3 Accounting Management

This functional area of network management is responsible for tracking the usage of network resources and services for accounting and billing purposes. Remarkably, accounting and billing information is fundamental to understanding usage patterns and foreseeing network growth and investments.

Functionalities

The accounting management area encompasses primarily the functionalities of usage data collection, billing and charging, resource allocation, quota management, and cost reporting, which are also useful to support other FCAPS-related tasks, such as resource usage optimization.

- Usage Data Collection. This functionality involves collecting data about the usage of network resources (e.g., bandwidth, storage, and processing power) by different users (residential or commercial) or services. For instance, bandwidth metering enables billing subscribers according to a defined data rate usage or a total data amount, charging additional fees if they exceed established thresholds. In turn, storage metering allows billing subscribers to consider the used storage capacity and charge extra when they exceed a maximum usage limit. Metering processing enables charging subscribers, taking into account the number of processors used from the total capacity rented.
- Billing and Charging. Billing is the process of invoicing customers or clients for the goods or services they have received. A charge is an expense item. An invoice is a bill that summarizes a group of incurred charges or future charges during a specific time period. In this regard, this functionality refers to calculating charges based on usage data collected and generating bills for subscribers or companies. Beyond billing based on metered data, subscribers can be differentiated by profile, for instance, between residential and commercial. Revenue tiers, into Bronze, Silver, and Gold, can also differentiate them. Additionally, this functionality can include support for real-time charging for various services, such as on-network gaming, VoIP, and cloud computing.
- Quota Management. This functionality relates to implementing quotas or usage limits for subscribers (residential or commercial, Bronze, Silver, or Gold) to prevent excessive resource consumption. Quota management supports the billing and charging functionality.
- Resource Usage Optimization. This functionality involves mainly resource management, which can be performed by considering usage patterns, and aims to ensure subscribers can access the resources they need while managing costs and profit. Resource management comprises various phases [1]: admission control, allocation, scheduling, and orchestration. Admission control decides which resource requests from subscribers can be accepted (or not). Acceptance is based on policies focused on increasing network providers' revenue, prioritizing services, and avoiding idle resources. Resource allocation quantifies the quantity of resources to allocate to maximize network operators' long-term economic benefits and fulfill, among others, the QoS/QoE demanded by subscribers. Resource scheduling refers to programming the time in which the network must allocate resources to, for example, minimize the total execution and operation time of the network services and ensure improved performance. Resource orchestration is responsible mainly for handling service chains, life cycle of services, and dynamically tuning assigned resources; for this phase is also primordial consider performance requirements and network status.
- Cost Reporting. This functionality embraces generating and providing reports on network usage and costs. These reports are crucial in enabling network practitioners to make informed decisions regarding resource allocation, budgeting, and overall network planning.

2.1.4 Performance Management

This functional area of network management involves monitoring, analyzing, and optimizing the performance of network components and services to ensure that they meet QoS/QoE and SLAs.

Functionalities

The Performance Management embraces primarily the following functionalities: performance monitoring, performance analysis, capacity planning, optimization, and SLA/XLA handling.

- Monitoring. This functionality enables the tracking and collection of performance metrics across the network, including latency, throughput, packet loss, and error rates. The collected data serves as input for the performance analysis functionality. Performance monitoring is based on various well-known network protocols. Some of them are described below.

  - SNMP is the most popular protocol in network management monitoring. By SNMP, network practitioners can query network devices (e.g., routers, switches, and servers) via GetRequest, GetNextRequest, and GetBulkRequest messages to retrieve performance metrics (e.g., bandwidth usage, CPU load, memory usage, and uptime). Moreover, SNMP can trigger alerts when setup thresholds are surpassed via Trap and InformRequest messages.
  - NetFlow is a network protocol introduced by Cisco to collect data about IP traffic. By this protocol, network administrators can, for example, obtain the source and destination IP addresses, the amount of data being transferred, and the duration of communication between endpoints.
  - Sampled Flow (sFlow), unlike NetFlow, which captures all traffic data, uses sampling (i.e., the process of selecting only a subset of packets from the total monitoring traffic) to capture packets and export flow data. Thanks to its sampling-based collecting feature, this protocol is often suitable for large-scale networks where full packet capture is impractical.
  - OpenFlow is a protocol used in SDN as a southbound interface to control and monitor network traffic. The monitoring capabilities offered by OpenFlow enable the collection of statistics at the switch, port, and flow table levels.
  - The Internet Control Message Protocol (ICMP) is helpful in measuring network latency and packet loss, providing insights about a network path's performance. This protocol serves as a base for networking tools, including ping and trace-route.

- Analysis. This functionality involves analyzing collected data on network and service performance to identify trends, peak and average loads, bottlenecks, and potential issues that could affect QoS/QoE and consequently, SLA/XLA compliance. Big Data enables performance analysis by facilitating the collecting, storing, analyzing, and interpreting the vast amounts of data generated by networks and their services. An accurate Big Data analytics depends on the quality of the collected data; therefore, it is necessary to ensure that the data is clean, accurate, and relevant to achieve reliable performance management. In Big Data, (i) Frameworks, such as Apache Flume and Apache Kafka, allow the collection, aggregation, and movement of large amounts of log network data in real time. (ii) Frameworks, such as Hadoop, Apache Spark, and NoSQL databases, can store large amounts of network data that grow over time in a distributed fashion to ensure redundancy, high availability, and quick access. (iii) Frameworks, such as Apache Kafka and Apache Storm, can process and analyze historical network data to perform descriptive (to discover historical pattern data) analysis. ML techniques can be combined with Big Data for descriptive, predictive (when and where performance networking issues are likely to occur), and prescriptive (what actions are necessary to perform to avoid network performance issues) analytics. Predictive and prescriptive analytics enable proactive network management, but handling vast amounts of network data raises concerns about data privacy and security. (iv) Frameworks such as Splunk, Grafana, and Kibana enable the visualization of network data analytics via simple, customizable, and powerful dashboards.
- Capacity Planning. This functionality focuses on generating a capacity plan. This plan primarily includes information on allocating resources to meet current and future demands; the functionalities of monitoring and performance analysis, as described above, support its construction. A plan capacity also contains scalability and budgeting details. Scalability refers to ensuring that the network can grow in response to business changes, and budgeting relates to the cost of upgrading the network. A trade-off between capacity to scale and costs is fundamental to getting a good capacity plan. Once a capacity plan is built, it must be implemented and monitored to ensure continuous tuning, as the business and technologies are highly dynamic.
- Optimization. This functionality implies implementing changes to improve the network performance and involves techniques such as ① traffic shaping, which aims to ensure sufficient network bandwidth is assigned

for critical applications framed into, for instance, Tactile Internet. ② Bandwidth throttling regulates traffic flow into a network, and ③ Rate limiting controls the outward traffic flow. ④ Load balancing consists of distributing network traffic across multiple links or servers to avoid congestion and ensure resources are not overwhelmed. ⑤ Minimizing the number of hops decreases the round-trip time of packets, and ⑥ edge processing assists in latency reduction since data are processed closer to their source. ⑦ Traffic prioritization helps ensure high-priority applications maintain performance even when the network is under heavy load. Redundancy and forward error correction help in reducing packet loss. ⑧ Network function virtualization enables the allocation of resources according to demand. ⑨ Software-defined networking permits the programmatic tuning of network control and configuration. ⑩ Green networking is oriented to control and reduce energy consumption to reduce the impact of network operations. ⑪ Cloud-based architectures help to scale as demand grows. ⑫ Continuous monitoring and feedback support the iterative and dynamic optimization of the network.

- SLA and XLA Handling. This functionality is conceived to ensure that the network and its services meet the performance requirements outlined in an SLA or XLA. An SLA is a formal contract between a network provider and a residential or commercial subscriber. This contract specifies the service expected level with measurable technical metrics (e.g., availability and reliability). A network provider usually suffers economic losses (e.g., financial penalties or free license extensions) with non-SLA compliance. Unlike SLA, an XLA focuses on end-user experience rather than technical metrics; remarkably, it incorporates non-technical metrics, such as user satisfaction, ease of use, and the overall impact on business outcomes.

2.1.5 Security Management

This functional area of network management aims to ensure the confidentiality, integrity, and availability of the network (including devices, data, and services) at the edge and inside the perimeter.

Functionalities

Overall, security management involves the following pivotal functionalities.

- The risk assessment and identification functionality comprises asset and threat identification, vulnerability assessment, and risk evaluation. The identification functionality refers to tagging all network assets (e.g., routers and data) that require protection. The threat identification functionality involves recognizing potential threats that compromise the network. The function of vulnerability assessment entails discovering and documenting network vulnerabilities. The risk evaluation prioritizes risks by measuring the potential impact of threats.
- The network security design and policy development functionality embraces security policy definition, network segmentation, security control implementation, authentication, and authorization. The definition of security policies involves formalizing a security strategy. A strategy can include rules (e.g., access control, firewall, and encryption) and incident response protocols, which provide clear guidelines on how to secure the network. Network segmentation refers to limiting access to different parts of the network to minimize the spread of an attack. The implementation of security control includes mechanisms and technologies to enforce security policies. Authentication involves users providing information about who they are (e.g., credentials that affirm they are who they claim), and authorization is about granting privileges to a user to access certain areas of the network or one of its components.
- The continuous monitoring and detection functionality encompasses monitoring, threat intelligence, alert generation, and preliminary incident response. The function of monitoring refers to detecting security incidents as soon as possible. The threat intelligence functionality aims to stay ahead of new and evolving security threats. The handling of automated alerts involves generating alerts when unusual or suspicious activity is detected. The function of preliminary incident response involves running predefined countermeasures to minimize damage.
- The incident response and mitigation functionality comprises various functionalities, namely incident detection, containment, eradication, recovery, and post-incident analysis. The incident detection involves confirming the

occurrence of an incident and assessing its impact. The functionality of containment isolates the attack. Isolation is used to prevent an attack from spreading within the network and consequently minimize the damage. The eradication function is designed to eliminate the threat and ensure that it cannot re-enter the network. The functionality of recovery involves restoring affected systems and bringing them back online in a secure state. The post-incident analysis is conducted to understand what happened, how the attack occurred, and how the response could be improved.

- The patch management and updates functionality encompasses mainly functionalities related to vulnerability patching and software and hardware upgrades. The vulnerability patching aims to mitigate security risks associated with known vulnerabilities by keeping operating systems up to date. Furthemore, upgrading software and hardware is essential for achieving a modern, resilient network that is resilent to attacks; in this regard, outdated software and hardware should be upgraded to ensure compatibility with current security standards.
- The compliance and auditing functionality includes security audits to evaluate the effectiveness of the network's security policies and guarantee compliance with legal and industry standards, such as ISO/IEC 27002 and the General Data Protection Regulation (GDPR). ISO/IEC 27002 provides guidance and best practices that organizations can follow to design, implement, and improve a cybersecurity-oriented information security management system. In turn, GDPR is a European law for data treatment. Additionally, this functionality includes, for the sake of transparency, recording and reporting the current and past network security status.
- The continuous improvement and optimization of network security encompasses various functionalities, including regular training, proactive threat hunting, and optimization of security tools.

  - Training aims to reduce human error. This reduction can be achieved by providing regular security training for employees, ensuring they understand security policies and recognize potential threats. The ultimate goal is that they know how to respond to security incidents.
  - Proactive threat hunting refers to building the organization's capacity to detect network threats before they cause damage.
  - The optimization of (software and social) security tools requires periodic review to evaluate whether they are working efficiently and can handle evolving threats; otherwise, they must be optimized or updated.

2.2 Cognitive Network Management

Cognitive network management can be perceived as a natural evolution in realizing the functional areas of network management, which leverages AI, ML, and cognitive computing to enable networks that can manage themselves autonomously. The cognitive network management architecture, as illustrated in Fig. 2.2, primarily consists of sensors, actuators, a closed or open control loop (see Sect. 2.2.1), self-management functions (see Sect. 2.2.2), and network management policies (see Sect. 2.2.3). From a broad perspective, the sensors and actuators interface the physical infrastructure of the network and the cognitive network management system. Sensors gather real-time raw data about the network's status. Actuators physically or logically apply actions to change the network behavior.

2.2.1 Cognitive Loop

A cognitive control loop involves a set of components and operations devised to support the creation of networks with self-monitoring, self-configuration, self-healing, self-optimization, and self-protection capabilities; remarkably, if involving a cognitive and closed control loop, the network management functional areas can be matched to self-healing (faults), self-configuration (configuration), self-monitoring (accounting), self-optimization (performance), and self-protection (security). IBM introduced the Monitor, Analyze, Plan, Execute, and Knowledge (MAPE-K) control loop framework for autonomic computing in about 2003, which has been widely adopted for cognitive network management. The MAPE-K loop provides a structured, iterative approach that enables networks to adjust their behavior automatically based on real-time data.

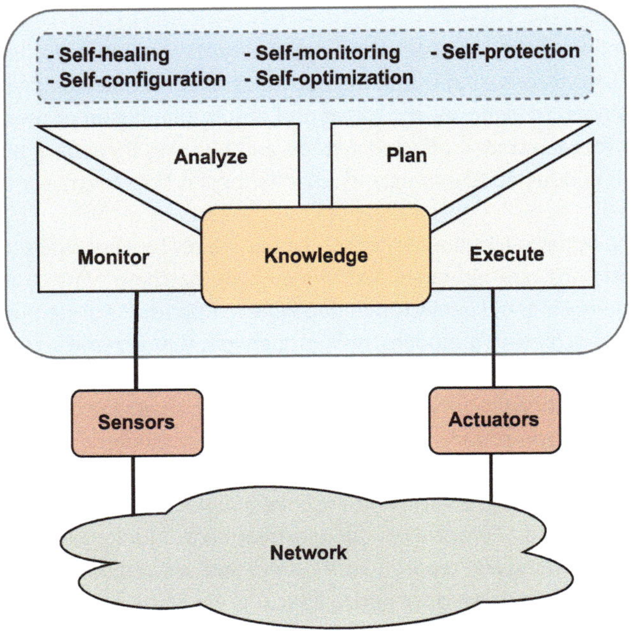

Fig. 2.2 Cognitive network management

At a high level of abstraction, the cognitive control loop operates continuously and cyclically, as follows. ① In the monitor phase, real-time data from the network is continuously gathered. ② The analysis phase refers to processing the collected data to identify patterns, issues, trends, or optimization opportunities in the network. ③ In the planning phase, an actionable plan to face issues or realize optimizations is generated. ④ The execution phase is in charge of implementing and deploying the action plan in the network. ⑤ The knowledge phase comprises storing the results and learned lessons from the results of previous phases for future decision-making.

Phases

- Monitor. This phase, which encompasses all network management functional areas, aims to collect raw data from all hardware and software within the networks. As mentioned earlier in the FCAPS description, several protocols facilitate this phase, including SNMP, NetFlow, and sFlow. The gathered data serves as input for the subsequent stages, providing a comprehensive picture of the network's current state. Note that the gathering can be cognitive; for instance, an ML technique can be used to learn the suitable frequency for sampling data.
- Analyze. This phase aims to process the raw data collected during the monitoring phase to detect and predict issues related to the FCAPS. Big data (e.g., Hadoop), AI (e.g., Fuzzy logic), and (supervised, semi-supervised, and self-supervised) ML techniques are pivotal to analyze the high-dimensional data generated by (small, medium, and large) networks. In this vein, the analysis phase produces actionable insights to enhance the network, for instance, regarding performance and security.
- Plan. This phase aims to create a plan that outlines the best actions (immediate or long-term) to address identified issues or capitalize on the opportunities found in the analysis phase. The plan can be built with AI techniques like Automated Planning and/or ML techniques like DRL. Automated Planning allows creating plans (a set of possible actions) to go from an initial state (e.g., a congested network) to a goal state (e.g., a balanced network). DRL learns by interacting with the environment through a set of episodes in which an agent carries out actions that change the environment state and receives rewards. The objective is to determine the optimal action in each state that enables the optimization of a goal. The goal is to optimize resource usage in a data center network. It is also important to note that the mentioned techniques can be combined for plan generation. For instance, when framed within model-based DRL, a planning technique like Hierarchical Task Networking utilizes the experience obtained by the agent to model the environment and use this model

for updating the value function without interacting with the environment, thereby speeding up the learning process of a Q-learning agent.

- Execute. This phase aims to automatically deploy the plan built in the planning phase, ensuring the rapid and efficient implementation of network changes related to FCAPS. Some of the networking paradigms that facilitate plan execution are NFV and SDN. NFV is fundamental in executing tasks associated with re-allocating networking and computational resources in cloud networking. SDN is pivotal to network management automation; for instance, OpenFlow (one of the most widely used protocols for SDN deployment) helps change flow tables to modify packets' routes. Orchestrator technologies, such as Kubernetes and Docker Compose, are also essential to coordinate the execution phase. Kubernetes and Docker Compose enable communication, scalability, security, and external access to containerized applications. Once the plan is executed, it is necessary to observe the network behavior to evaluate the results.
- Knowledge. This phase primarily involves the knowledge repository conceived to ensure the network management system learns from experience, improving its performance and adaptability over time, and allowing the network to become more autonomous and intelligent. The knowledge base generates knowledge from the information generated by the other phases and provides feedback to the subsequent execution. Pivotal tools for this phase are ontologies, graph databases, and ML. Here, ontologies (i.e., a formal, structured representation of knowledge that define the entities, relationships, and rules within a domain, as networking is) enrich semantically the knowledge base, for example, they describe what a "router," "flow," or "QoS policy" is and how they relate, enabling reasoning and interoperability across heterogeneous management systems. Graph databases are storage systems that represent data as nodes (entities) and edges (relationships), which are optimal for querying and traversing complex relationships rather than simple tabular data. A graph database can support queries such as "find the shortest path across congested links" or "identify dependencies between services", helping, for instance, the Analyze and Plan phases. ML enriches the knowledge base with learned models, updated patterns, and feedback loops.

2.2.2 Self-management Functions

In autonomous cognitive network management, self-healing (e.g., a network automatically reroutes traffic when a router fails), self-configuration (e.g., a switch is automatically configured when added to the forwarding plane of an SDN), self-monitoring (the network alerts administrators when computational resources usage exceeds a threshold in an NFV environment), self-optimization (e.g., softwarized network redistributes traffic to reduce congestion), and self-protection (e.g., the network blocks suspicious IP addresses to prevent a Distributed Denial of Service—DDoS, a cyberattack multiple compromised systems flood a target server, service, or link with massive volumes of traffic or resource requests for overwhelming its capacity and making it unavailable) are network management functions devised to enable a network to autonomously manage its operations, improve its performance, and protect itself from threats. Self-management functions are essential in novel and intelligent network management systems to reduce human intervention, achieve resilience, and operate the network (including hardware, software, and firmware) smoothly and securely in dynamic environments.

Self-healing

This self-management function refers to the ability of a network to detect, diagnose, isolate, notify, resolve, document, and automatically restore from errors, faults, fails, or failures without human intervention. In this function, the monitor continuously collects raw data about the network devices and services; the analyzer detects when issues arise by processing the raw data; the planner defines corrective actions (e.g., isolate a router and put functioning its backup); the executor realizes the actions in the hardware and software involved in the issues to restore the network to normal operation, and the knowledge base stores all issue and solution related info and learned lessons. The fault management functionalities must operate in a closed control loop in self-healing. The loop needs AI and ML techniques to enhance availability and network resilience. Considering the self-healing functionalities and the cognitive architecture:

- The Monitoring and Analysis phases embrace the functionalities of detection and diagnosis.
- The Planning and Execution phases encompass the functionalities of isolation, notification, resolution, and restoration.
- Knowledge Base embraces documentation and learning.

Self-configuration

This self-management function refers to the ability of a network to take inventory, handle changes, versioning, and automatically provision configurations when new devices or services are added or changes occur, all without human intervention. In this function, the monitor and analyzer detect hardware, software, or firmware change; the planner defines the configuration to apply and the policies to review to ensure the new element (e.g., router's firmware) complies performance rules; the executor provisions the configuration in the involved element; and the knowledge base stores all configuration and policies related info and learned lessons. In self-configuration, the configuration management functionalities must operate in a closed control loop. The loop requires AI and ML techniques to minimize human error and expedite the deployment of new devices and network services. From a broad perspective and considering a cognitive system, for achieving self-configuration:

- The Monitoring and Analysis phases embrace the functionalities of inventory and handle changes.
- The Planning and Execution phases encompass the functionality of provisioning.
- The Knowledge Base spans the functionalities of inventory and version control.

Self-monitoring

Self-monitoring is pivotal to achieving the other self-management functions and refers to the ability of the network to continuously observe its own performance, health, and security status without human intervention. This function must operate in a closed control loop to enable early problem detection and proactive responses. AI and ML techniques must be utilized to obtain real-time data, enhance data quality, minimize monitoring traffic overhead, and reduce human involvement. Relevant data quality aspects to consider are noise (irrelevant variations that obscure patterns in data), missing values (absence of required data samples that can hinder analysis), outliers (unusual values), dimensionality (a high number of features can increase complexity), inconsistency (contradictory information across data samples), redundancy (same data that does not add new information), data volume (challenge storage and processing), heterogeneity (diversity of data formats, sources, or structures that must be integrated), and timeliness (degree to which data is up to date and available when needed). Monitoring traffic overhead refers to the traffic generated to collect data on device and network services. Remarkably, human involvement is prone to error and, therefore, should be avoided.

Self-optimization

This self-management function refers to the network's ability to optimize profit while maximizing performance and minimizing resource utilization over time by automatically adjusting device and service configurations based on real-time conditions, without requiring human intervention. The performance management functionalities must operate in a closed control loop in self-optimization. The loop requires AI and ML techniques to enhance the effectiveness of the network (devices and services), including QoS/QoE achievement and SLA/XLA satisfaction, while reducing the need for manual tuning. From the perspective of a cognitive control loop:

- The Monitoring and Analysis phases embrace the functionalities of performance monitoring and performance analysis.

- The Planning and Execution phases encompass the functionalities of capacity planning and optimization.
- Knowledge base embraces documentation (capacity planing) and serves as foundation to optimize the making decisions.

Self-protection

This self-management function refers to the network's ability to continuously apply security policies to detect and defend against security threats in real time, preventing potential attacks or unauthorized access without manual intervention. The security management functionalities must operate in a closed control loop in self-protection. The loop needs AI and ML techniques to achieve real-time protection against attacks, reduce the risk of breaches, and ensure data privacy, integrity, and confidentiality. From the cognitive control loop perspective:

- The Monitoring and Analysis phases embrace the functionalities of continuous monitoring, risk assessment, and security threats identification.
- The Planning and Execution phases encompass the functionalities of network security design, security policy development, incident response and mitigation, patch and security updates handling, and network policies and tools auditing to comply with regularization and standards.
- Knowledge Base embraces continuous learning, improvement and optimization of network security, including data, devices, and services.

2.2.3 Network Management Policies

Policies, defined by network practitioners, guide the behavior of cognitive network management systems. Network management policies often follow the structure IF (condition), THEN (action) to specify performance objectives and compliance with regulatory requirements. As network management policies (e.g., tactile internet services must operate with an end-to-end delay no higher than 5 ms) are defined at a high-abstraction level, they must be translated to actionable (e.g., configuration commands) commands by network elements; to achieve such a translation, it is pivotal to combine AI and ML techniques.

Network management policies are typically expressed using natural language or specialized languages that allow network practitioners to define rules, actions, and conditions for managing network (physical or virtual) devices and traffic. The following is a non-exhaustive list of policy languages.

- Policy Description Language (PDL) allows network practitioners to express policies (conditions, actions, and preferences for managing resources) in a human-friendly format. Listing 2.1 exemplifies a policy described in PDL that prioritizes real-time video traffic over best-effort traffic during congestion.

```
POLICY "Video Traffic Prioritization"
WHEN
    Network congestion level > 80% utilization
AND
    Application type = "Video Streaming"
THEN
    Allocate minimum bandwidth = 5 Mbps
    Set priority = HIGH
    Apply queue = RealTimeQueue
PREFERENCE
    Prefer paths with lower delay (< 30 ms) over higher bandwidth

```

Listing 2.1 Example with policy description language

- Ponder. This language allows network practitioners to express obligation policies for event-triggered actions and authorization policies for access control.
- Path-Based Policy Language. This declarative language allows network practitioners to specify access control policies based on the paths in cloud and virtualized networks.
- XPath and XQuery. These languages enable network administrators to query and manipulate YANG configurations. YANG models define the structure and management of data in devices, while policies can be enforced using NETCONF commands.

Policies from the Self-management Functions Perspective

Network management policies provide guidance, constraints, and permissible actions for self-management functions, ensuring they operate within predefined rules and maintain consistency, compliance, and alignment with organizational goals.

- Self-healing policies are defined to ensure that the network automatically addresses errors, fails, faults, and failures, thereby improving availability and enhancing reliability. E.g., P1: <if a link fails in a softwarized and virtualized network, compute and install new rules in the flow tables to reroute traffic within 5 s to maintain connectivity>. P2: <move 5G enhanced mobile broadband services automatically to a user plane function with a higher capacity if the throughput by the current user plane function is lower than 100 Mbps>.
- Self-configuration policies aim to dictate how the network should automatically adjust its setup in response to its dynamic nature, such as new devices, network topology adjustments, and changes in traffic patterns. E.g., P1: <if an OpenFlow-enabled switch is added, automatically configure the security policies and rules in the flow table>. P2: <if an OpenFlow-enabled switch is added, automatically update the network topology vision>.
- Self-monitoring policies aim to enable the network to enhance its monitoring tasks based on its status. E.g., P1: <if network congestion is detected, prioritize business-critical applications and reduce bandwidth for monitoring tasks>. P2: <if network state is not changing, decrease the data collection frequency>.
- Self-optimization policies. These policies intend to guarantee that the network proactively meets SLA and XLA by governing resource management, including admission, allocation, orchestration, and scheduling. E.g., P1: <ensure a minimum 1 Gbps bandwidth for 6G slices of remote surgery>. P2: <if the end-to-end delay is higher than 10 ms, reroute the traffic of 5G ultra-reliable low-latency communication services>.
- Self-protection policies aim to assist the cognitive network management system in proactively defending the network against internal and external cyberattacks, unauthorized access, and data breaches by automatically enforcing security controls. P1: <block traffic coming from non-well-known IP addresses attempting to access internal database servers>. P2: <encrypt all sensitive data transmitted between geographically separated campus networks belonging to the same autonomous system>.

Policies in the Closed Control Loop

Network management policies offer guidance, constraints, and permissible actions for the MAPE-K loop to ensure the cognitive network management system operates within predefined rules, maintaining consistency, compliance, and alignment with organizational goals.

- Monitoring Phase. Policies in this phase help define what and when data must be collected (e.g., bandwidth, latency, security events) and establish thresholds that trigger actions when exceeded. E.g., P: <for non-critical services collect performance data every 180 s and for critical services collect data every 15 s>.

- Analysis Phase. Policies are defined in this phase to aid the cognitive network management system in interpreting the collected data, for instance, by determining what constitutes normal behavior, an anomaly, or a violation. E.g., P: <raise an alert if critical services are requesting more than 80% of available bandwidth>.
- Planning Phase. Policies in this phase are defined to guide the decision-making process, leading to the creation of a plan that outlines permissible actions, such as resolving issues or optimizing performance. E.g., P: <if critical services need more than 80% of available bandwidth, allocate it from non-essential services>.
- Execution Phase. Policies are defined in this phase to ensure that the planned actions are executable and comply with organizational goals and international standards. E.g., P: <when taking bandwidth from non-essential applications, avoid service outage on them>.
- Knowledge Base. Here, policies are defined to assist the cognitive network management system to learn from past actions by providing context on why certain decisions were made and how effective they were. E.g., P: <if a DDoS attack occurs, use the last top-k successfully applied plans>.

2.3 Zero-Touch Network and Service Management

Zero-Touch Network and Service Management (ZSM) is a management paradigm introduced by the European Telecommunications Standards Institute (ETSI) that refers to the fully automated deployment of network services, as well as their configuration, operation, and maintenance, without (or with minimal) human intervention. ZSM aims to create a network with self-management functions (i.e., self-monitor, self-configure, self-optimize, self-protect, and self-heal) to reduce operational expenditures and complexity of network management while improving the speed and efficiency of service delivery.

2.3.1 Architecture—Overview

From a broad perspective, the ZSM architecture illustrated in Fig. 2.3 comprises various components, including management domains, an inter-domain integration fabric, and end-to-end service management domains. These components focus on fully automating the deployment, configuration, and maintenance of network services to reduce operational and capital expenditures, facilitate faster service delivery, and enhance network reliability. Nowadays, implementing an AI/ML-based ZSM architecture to operate a complete network represents a tremendous challenge in the network management domain.

2.3.2 Management Domains

A management domain is a logical grouping of network functions, resources, or services managed autonomously. In this vein, types of management domains include Network Management Domain, Service Management Domain, Resource Management Domain, and Slice Management Domain. A network management domain refers to the process of performing self-operations over a network infrastructure (e.g., an OpenFlow-enabled switch, 5GCN, and 5GRAN). A service management domain manages services and applications to ensure they are deployed, monitored, optimized, and terminated as needed. A resource management domain refers to the management of physical and virtual resources (e.g., compute, storage, and network) required to support services and network functions. A slice management domain handles 5G network slices, including their creation, monitoring, and scaling. Remarkably, each management domain includes self-properties for network, service, and resource management, as well as an intra-domain integration fabric, to ensure that the self-management functions can interact and work together seamlessly. Furthermore, each can operate independently and interact with other domains through standardized interfaces.

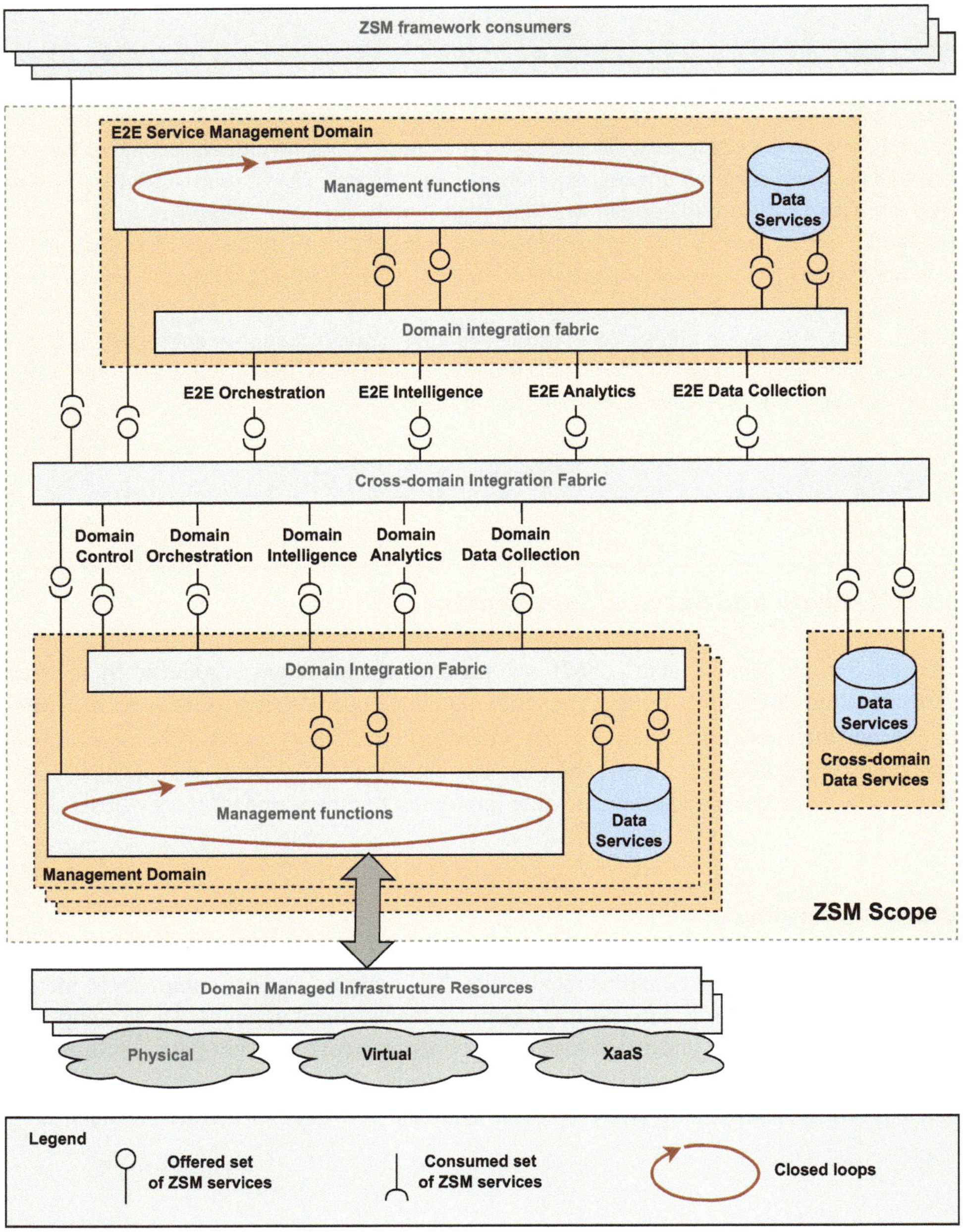

Fig. 2.3 ZSM architecture

A management domain includes:

- Self-management functions in ZSM encompass diverse self-properties, including self-monitoring, self-configuration, self-optimization, self-protection, and self-healing. A key foundation for realizing self-properties is the concept of closed-control-loop automation. Such a loop enables continuous monitoring of network status, analysis of performance data, informed decision-making, and prompt action-taking without human intervention (see Sect. 2.2). In ZSM, the control loop that supports the self-management functions is based on AI/ML techniques, and its behavior is guided by intents (see Sect. 2.3.5).
- Data Services refer to mechanisms and platforms responsible for collecting, processing, storing, and analyzing data within a specific management domain. These services extend beyond Domain Data Collection, which collects raw

data (e.g., traffic logs from routers) within a particular management domain. They focus on transforming raw data into actionable insights and making data available for use in decision-making, automation, and cross-domain interactions.

- The intra-domain integration fabric ensures that all elements within a management domain can communicate, share data, and coordinate actions effectively, enabling automation and coordination. In this regard, the intra-domain integration fabric orchestrates self-management functions to ensure they can jointly operate effectively and coordinately. This integration fabric also enables data exchange among self-management functions and between these functions and the managed network (including services and resources), offering a real-time messaging layer crucial for executing management services correctly and promptly. Moreover, this fabric is pivotal to enforcing management policies in a network management domain and permits integrating data services with automation and intelligence capabilities at the intra-domain level. Remarkably, such an integration paves the way for autonomous addressing of network events via data-driven decision-making.

2.3.3 Inter-domain Integration Fabric

The integration fabrics provide and consume management services by following either the request-response or request-response-notify model for service invocation or the publish-subscribe model for live event streaming. These services, which are necessary for closed-loop automation at the management domain level, are assembled into logical groups: Domain Control, Domain Orchestration, Domain Intelligence, Domain Analytics, and Domain Data Collection.

- Domain Control aims to manage and enforce intra-domain policies for ensuring each domain operates autonomously and efficiently in a decentralized manner. Decentralization allows networks and services to scale and adapt dynamically. Moreover, a domain control must interact with other domain controls and the E2E service management domain function. This interaction guarantees that the network and service infrastructure operate cohesively to meet end-to-end service goals and accomplish global business objectives.
- Domain Orchestration represents an orchestrator that coordinates and automates repetitive tasks within a particular management domain. In this regard, it manages the life cycle of resources (e.g., memory, CPU, and bandwidth) and services within the domain, ensuring their correct and timely provisioning, optimization, scaling, and termination. The orchestrator is also responsible for enforcing domain-level policies to ensure that the management domain operates according to rules and performance standards defined in SLAs. Beyond its local operation, it must coordinate with the E2E orchestrator to accomplish end-to-end QoS/QoE requirements of services and global business goals.
- Domain Data Collection aims to obtain raw data from network components, services, and resources within a specific management domain. Collecting (high-quality) data is crucial for enabling automation, informed decision-making, and optimization within the domain, as well as for sharing information across inter-domain boundaries to support end-to-end service management. Beyond involving data sources (e.g., logs containing values of performance metrics for routers, links, and switches), the collection also consists of handling. (i) Filtering since not all collected data is valid. (ii) Depending on the management task, aggregate data at the device, service, or network segment level. (iii) Storing data in a centralized or decentralized way to process it in a near or real-time way. The domain data collection interacts as follows. Its data can be fed into various domains, including orchestration, analytics, and intelligence, to support decision-making. Domain data can be shared via the inter-domain integration fabric to enable the aggregation and analysis of data from multiple domains.
- Domain Analytics refers to analyzing data collected within a specific management domain to gain insights, support decision-making, and enhance the automation of network and service management tasks. AI and ML techniques permit the development of domain analytics with descriptive, predictive, and prescriptive scopes. Descriptive analytics is used to answer the question, "What happened in a management domain?" by identifying trends and relationships

within current and historical data. By predictive analytics, data are used to forecast future trends and events, which can help inform strategic decisions and answer the question, "What is going to happen in a management domain?" Prescriptive analytics performs a step further than descriptive and predictive analytics to enable the proposal of optimal actions to deploy in a management domain. The domain analytics interacts as follows. It feeds insights into the domain orchestrator to enable knowledge-based decisions. Domain analytics not only analyzes data locally within the domain but also shares relevant insights via the inter-domain integration fabric to support end-to-end service management. Domain analytics relies on the continuous flow of data offered by domain data collection.

- Domain Intelligence copes with the integration of advanced AI and ML techniques within a specific management domain to enable autonomous decision-making, optimization, and adaptive behavior. Notably, the intelligence allows any management domain to learn from data (i.e., data-driven), predict future events (i.e., predictive behavior), and make real-time adjustments (i.e., prescriptive behavior) to network and service management without human intervention; the predictive and prescriptive behaviors enable proactive management and adaptation to network dynamics. Domain intelligence interacts as follows. It depends on domain analytics for processing and interpreting the data collected within the domain. Domain intelligence works closely with the domain orchestrator to provide, for instance, data-driven insights and recommendations about management actions. Domain intelligence operates within its domain but may collaborate with intelligence systems in other domains through cross-domain coordination.

The integration fabrics at the E2E service management domain level also provide and consume management services, following either the request-response or request-response-notify model for service invocation, or the publish-subscribe model for live event streaming. At this level, the management services are logically grouped into E2E Orchestration, E2E Intelligence, E2E Analytics, and E2E Data Collection.

- E2E Orchestration refers to synchronizing automatically or autonomously (reducing human intervention, leading to response times improvement) resources and services across multiple management domains to ensure services are delivered efficiently and meet SLAs. The E2E orchestrator is also responsible for managing the entire lifecycle of services (from provisioning and scaling to monitoring and decommissioning) across all involved domains. Moreover, E2E orchestration coordinates actions involving diverse management domains to ensure that each domain's resources and services are aligned with the overall service and business objectives. E2E orchestration utilizes functionalities provided by E2E data collection, E2E data analytics, and E2E intelligence to execute the necessary actions across multiple domains, thereby realizing service delivery as demanded.
- E2E Data Collection gathers data from various sources across different management domains to create a unified dataset representing the entire service delivery chain. This data enables cross-domain management and provides a comprehensive view of the network and its services. Data includes real-time metrics and historical data to pave the path for descriptive, predictive, and prescriptive behavior. Regarding E2E data collection, it is also remarkable that data is often stored in a centralized or federated repository accessible for E2E analytics (primarily), E2E orchestration, and E2E intelligence.
- E2E Analytics aims to analyze data collected from multiple domains to generate insights that drive autonomous decision-making and optimization for end-to-end services and the network as a whole. By performing descriptive, predictive, and prescriptive analysis, this ZSM component can provide a holistic view of the network and its services, enabling an understanding of their current and past functioning, prediction of their future behavior, and informed tuning decisions. E2E Analytics interacts as follows: it relies on E2E data collection to manage the lifecycle of end-to-end services effectively. Moreover, insights produced by E2E analytics are used by E2E orchestration and E2E intelligence.
- E2E Intelligence empowers the ZSM architecture to make informed decisions, optimize operations, and adapt autonomously to changing conditions. In this regard, E2E Intelligence leverages E2E data collection and E2E analytics to offer descriptive, predictive, and prescriptive insights to the entire network and its end-to-end services, with

or without minimal human intervention. The E2E intelligence service must learn from streaming and continuously from the available data and analysis to improve its decision-making capabilities, making the network and services more self-driving. Remarkably, E2E Intelligence is designed to enhance orchestration and analytics by leveraging AI and ML techniques to achieve more accurate diagnoses, predictions, and prescriptions, automate decisions, and optimize service delivery.

The inter-domain integration fabric also interacts with cross-domain data services to perform its functions. These services extend the concept of data services beyond a single domain to enable data storage, sharing, and analysis across E2E service management domains. In this regard, cross-domain data services provide data exchange across domains, global data aggregation, capabilities for centralized data analytics, and enable models to learn from data collected across the network and service landscape. Cross-domain data services are essential to achieve a global vision of the entire network and its services and, further, play a crucial role in the way to accomplish end-to-end service management by ensuring that data relevant to the whole service lifecycle is always accessible to the domains responsible for different segments of the service.

2.3.4 E2E Service Management Domains

An E2E service management domain aims to manage and orchestrate services across management domains to ensure services are delivered, maintained, and optimized across diverse networks or network segments. In particular, this domain uses its components (i.e., self-management functions, intra-domain integration fabric, and data services) and interacts via inter-domain integration fabric with E2E Services, Domain Services, and Management Domains to perform various global tasks, including: (i) it manages the entire lifecycle of a service, including service creation, scaling, monitoring, optimization, and decommissioning (i.e., a critical process that involves removing a system from service, securing its data, and disposing of its hardware and software). (ii) It operates as a central orchestrator to guarantee that services are provisioned, configured, monitored, and maintained across various domains. (iii) It ensures that QoS/QoE and SLA policies are applied and consistently met across all service delivery. (iv) It leverages automation and AI/ML techniques to realize self-management of services across domains in real-time and end-to-end.

The main interactions of the E2E service management domain and the other ZSM components are as follows. ① It cooperates with an individual management domain to guarantee that it carries out its role effectively in end-to-end service delivery; for services that span multiple domains, the E2E service management domain coordinates the actions of individual management domains to assure the service functions seamlessly across the entire network. ② It relies on cross-domain data and analytics to get a global view of the service's performance across all domains. ③ It oversees orchestration across domains to ensure services are consistently delivered according to QoE/QoE and SLA/XLA.

2.3.5 Intents

Intents are pivotal in ZSM. Intents enable ZSM since they do not require manual intervention, allowing the network to operate autonomously and fulfill business objectives with minimal human intervention. Furthermore, thanks to their declarative nature, intents are understandable for both technical and non-technical network practitioners. They can be understood as high-level, declarative expressions that describe networks or services' desired outcome, but it does not provide step-by-step instructions on achieving the outcome. Intents typically operate in an end-to-end fashion, which aligns with self-management functions and closed control loops.

An Intent-driven ZSM system operates as follows.

- Definition. Network practitioners (e.g., administrators, service managers, or business leaders) can define intents in natural language or in a high-level language (e.g., using JSON notation). E.g., I1: <ensure extremely low latency and ultra-high reliability for tactile internet services>. I2: <ensure ultra-high throughput for enhanced mobile broadband services>.
- Translation. The ZSM system translates the intent autonomously. In particular, the Domain Orchestrator or the E2E Orchestrator can perform this translation using AI and ML techniques, such as a Large Language Model (LLM), Small Language Model (SLM), or Fine-tuned LLM/SLM (FLM), breaking down the abstract intent into specific actions that the network can execute.
- Autonomous functioning. The ZSM system must ensure that the intent is executed without human intervention across the network and services, comprising one or more self-management functions that, for instance, apply the necessary changes to network configurations, resource allocation, or service parameters.
- Continuous monitoring and optimization. Once the intent is applied, the ZSM-driven network continuously monitors itself (self-monitoring to get, for instance, performance, service quality, and other vital metrics) to verify that it is achieving the desired outcome (goal). If the ZSM-driven network detects deviations from the intent-stated goal (e.g., higher-than-expected latency), it automatically executes self-management functions (e.g., self-optimization and self-configuration) to seek to meet the intent.
- The ZSM-driven network needs to leverage AI and ML techniques to learn from previous executions of self-management functions (involving actions and network statuses) and external factors to optimize future decisions and actions that lead to meeting intents. Remarkably, this continuous improvement process ensures that the network evolves to fulfill intents more effectively.

By autonomously translating business intents into service intents, service intents into resource intents, and resource intents into tasks actionable within the network, it is possible to achieve a ZSM-driven network.

Intents can be categorized into business intents, service intents, and resource intents [2, 3].

- Business intents help reflect the organization's strategic goals in the network infrastructure; non-technical network practitioners, such as business managers or service owners, typically define these. Therefore, this type of intent aligns network behavior with the overall business objectives to accomplish SLAs and XLAs. E.g., I1: <minimize operational costs while maximizing user experience>, help define goals centered on profit maximization and delivering excellent customer experience.
- Service intents help define the goals related to the level of QoS and QoE that the network needs to provide in its services; technical network practitioners, such as network administrators or cloud managers, typically define them. Note that in the ZSM domain, service intents must be oriented to meet the defined business intents. E.g., I1: <guarantee extremely high bandwidth for video streaming services>.
- Resource intents help define goals for managing and operating computational (memory, storage, and processing) and networking (bandwidth) resources; similar to service intents, technical network practitioners typically define resource-level intents. This type of intent aims to ensure that resources are allocated effectively and efficiently to support the accomplishment of the service intent. E.g., I1: <ensure redundancy across all network links passing video streaming to prevent service outage>.

2.4 Summary

This chapter presented an overview of network management paradigms. Initially, it introduced the main concepts related to functional areas of network management from the perspective of the International Organization for Standardization. Afterward, it discussed the primary foundations of cognitive network management architecture. Finally, it presented an architectural view of zero-touch network and service management.

Conceptual Questions

2.1 Define what network management is.

2.2 Describe the network management functional areas.

2.3 Discuss the primary functions of each network management area and relates to ML techniques.

2.4 Define what cognitive network management is.

2.5 Discuss cognitive network management components from a ML perspective.

2.6 Describe each self-management function and relates to ML.

2.7 What is the difference between a closed and an open control loop in cognitive network management.

2.8 What are the differences and similarities between FCAPS and self-management functions?

2.9 Define what network management policies are. Also, explain their role in the network management functional areas and self-management.

2.10 In a nutshell explain the zero-touch network and service management architecture, including primary goal and components.

2.11 Establish differences between network management, cognitive network management, and zero-touch network and service management.

2.12 Define which ZSM components can benefit from ML techniques. Also, discusses how those benefits could be carried out.

Solutions of This Chapter

2.1 Problem. Define what network management is.

Solution. Network management refers to the processes, technologies, and tools used to carry out FCAPS tasks that optimize computer or telecommunications networks regarding reliability, security, performance, and economic efficiency. Effective network management is critical for maintaining seamless communication, minimizing downtime, and supporting business operations.

2.2 Problem. Describe the network management functional areas.

Solution. Fault Management mainly involves detecting, diagnosing, and solving network problems. Configuration Management pivotally consists of setting up and managing network devices like routers, switches, and firewalls. Accounting

Management fundamentally involves tracking network usage for billing or resource allocation purposes. Security Management primarily points to protecting networks from unauthorized access, attacks, and vulnerabilities. Performance Management refers to optimizing computer and telecommunication networks in terms of QoS and QoE metrics and operators' profit.

2.3 Problem. Discuss the primary functions of each network management area and relates to ML techniques.

Solution. As fault management demands detecting, diagnosing, and overcoming network problems to minimize downtime, ML models can predict failures, identify root causes, and even prescript solution steps. Since configuration management involves setting up and managing network devices to ensure proper operation and policy compliance, ML models can automate configuration tasks and optimize settings based on traffic patterns or performance goals. Since accounting management targets tracking network usage for billing, resource allocation, or capacity planning, ML can analyze usage patterns to predict future demand, detect fraudulent activities, or optimize billing models. In performance Management, which at a high-abstraction level aims to optimize networks, ML can be used to predict traffic loads, compute optimal routes, and allocate resources dynamically. In Security Management, which seeks to protect the network from threats, unauthorized access, and vulnerabilities, ML can be used in intrusion detection systems, malware detection, and overall in identifying and mitigating security threats in real-time. On the other hand, since programmability management focuses on ensuring network controllers are updated with the latest patches and features in programmable networks, ML models can facilitate the prediction of the impact of updates, automation of patch management, and identification of vulnerabilities that need urgent attention. Remarkably, in 5G and beyond networks, ML models are pivotal in enabling autonomous optimization, which is, in turn, fundamental to handling the complexity of such networks.

2.4 Problem. Define what cognitive network management is.

Solution. Cognitive network management uses primarily ML and cognitive computing fundamentals and techniques to manage computer and telecommunication networks by realizing self-healing, self-configuration, self-monitoring, self-optimization, and self-protection. Each cognitive network management loop phase (monitor, analyze, plan, and execute) and the knowledge base can benefit from AI (e.g., automated planning and NLP), GAI (e.g., RAG and transformers), and ML (e.g., supervised, unsupervised, self-supervised, and semi-supervised) fields.

2.5 Problem. Discuss cognitive network management components from a ML perspective.

Solution. The monitoring phase can leverage, for instance, data cleaning, normalization, and feature extraction techniques to process the raw data collected and RL/DRL to tune collection frequency. The analysis phase can use, for example, supervised Learning for fault classification and traffic prediction, unsupervised Learning for anomaly detection and clustering of flows, and RL/DRL for dynamic resource allocation and optimization. The planning phase can benefit from RL/DRL to calculate optimal routes according to diverse criteria (e.g., less delay and packet loss or highest available bandwidth) and the number of virtual network functions or container network functions to instantiate to deploy a service with a demanded QoS. The execution phase can use rule-based systems to automate the execution of the task defined by planning. The knowledge base can leverage online learning algorithmsÂ to incorporate real-time network data to keep knowledge updated and RL/DRL to refine knowledge based on the outcomes of earlier actions.

2.6 Problem. Describe each self-management function and relates to ML.

Solution. From a wide perspective, when faults, configuration, accounting, performance, and security management are carried out by AI, GAI, and ML techniques, they become self-healing, self-configuration, self-monitoring, self-optimization, and self-protection.

2.7 Problem. What is the difference between a closed and an open control loop in cognitive network management.

Solution. A closed-loop control system continuously monitors the network, analyzes data, and automatically takes action based on feedback. It forms a "closed" cycle where the cognitive network management system learns from its actions and adapts over time. An open-loop control system does not automatically take action. It offers recommendations or suggestions to network practitioners (humans in the loop) responsible for making decisions that will be executed.

2.8 Problem. What are the differences and similarities between FCAPS and self-management functions?

Solution. FCAPS can be realized without involving AI, GAI, and ML. However, when the FCAPS functional management areas are aided by AI, GAI, and ML techniques, they become self-management functions.

2.9 Problem. Define what network management policies are. Also, explain their role in the network management functional areas and self-management.

Solution. Management policies are essential to ensure efficient, secure, and reliable operation of traditional and programmable networks. In self-managing networks, policies are enablers for autonomous and proactive decision-making, reducing the need for human intervention. In fault management and self-healing, policies establish how errors, faults, and failures in the network are detected. Policies govern how network devices are configured, updated, and maintained in configuration management and self-configuration. In accounting management, policies define how the network usage is tracked, billed, and allocated. In performance management and self-optimization, policies ensure the network meets performance requirements and optimizes resource usage. In security management and self-protection, policies help to define how to protect the network from threats and ensure compliance with security standards.

2.10 Problem. In a nutshell explain the zero-touch network and service management architecture, including primary goal and components.

Solution. ZSM architecture leverages AI, GAI, ML, and intent-based policies to realize self-management functions. It aims to simplify network operations, reduce costs, and improve efficiency, especially in 5G and beyond networks. From a high-abstraction level, ZSM includes management domains (a domain is a logical grouping of network management functions, resources, or services managed autonomously), integration fabrics (it provides and consumes management services by following either the request-response or request-response-notify model for service invocation or the publish-subscribe model for live event streaming), E2E service management domains (each one manages and orchestrates services across management domains to ensure services are delivered, maintained, and optimized across diverse networks or network segments), and intents (they are high-level, declarative expressions describing a network or service's desired state or outcome—usually in an end-to-end fashion, such as maintaining specific security or QoS level, which aligns with self-management functions and closed control loops).

2.11 Problem. Establish differences between network management, cognitive network management, and zero-touch network and service management.

Solution. Network management, cognitive network management, and ZSM were conceived to offer low to moderate, high, and whole automation levels. In this regard, the first, second, and third need significant, minimal, and non-human intervention related to the non- or limited use of basic and advanced AI/GAI/ML techniques. Therefore, traditional network management systems usually behave reactively, cognitive management systems offer predictive and proactive behavior, and ZSM systems seek to behave descriptively, predictive, and prescriptively to offer fully automated lifecycle management.

2.12 Problem. Define which ZSM components can benefit from ML techniques. Also, discusses how those benefits could be carried out.

Solution. ML techniques are ZSM enablers because they enable automation, predictive analytics, intelligent decision-making, and adaptive control. ZSM-enabled networks are fully autonomous, self-optimized, and operate with minimal human intervention. Pivotal ZSM components like management domains, E2E management domains, integration fabrics, and intents can leverage ML techniques in this path. The self-management functions can benefit from ML to, for instance, handle faults, setup and admin configurations, define pricing schemes, optimize network segments, and manage zero-day attacks autonomously.

References

1. J.A.H. Sánchez, O.M.C. Rendon, Deep reinforcement learning for resource management on network slicing: a survey. Sensors **22**(8), 3031 (2022)
2. A. Clemm, L. Ciavaglia, L.Z. Granville, J. Tantsura. Intent-based networking: concepts and definitions (2022). https://www.rfc-editor.org/info/rfc9315
3. J. Niemller, J. Silvander, P. Stjernholm, L. Angelin, U. Eriksson, Autonomous networks with multi-layer, intent-based operation: a truly autonomous network has the ability to change requirements dynamically without human involvement creating a network that can dothis requires the use of intents between multiple operational domains of the network architecture, including the business, service and resource layers. Ericsson Technol. Rev. **223**, 2–13 (2023)

Machine Learning in Networking: An Overview 3

3.1 Learning Paradigms

ML is part of AI, whose foundational concepts originate from contributions in diverse areas such as computer science, mathematics, philosophy, economics, neuroscience, psychology, and control theory [1]. In 1959, Arthur Samuel defined Machine Learning as *the field of study that allows computers to learn without being explicitly programmed*. Remarkably, ML goes beyond simply learning; the idea is to utilize and improve knowledge over time and with experience. From a high-abstraction perspective, ML techniques can be grouped into various paradigms based on how *learning* is achieved, namely *supervised*, *unsupervised*, *semi-supervised*, *self-supervised*, *deep*, *reforced*, and *generative*.

Supervised Learning utilizes labeled training datasets to create models that map inputs to their corresponding outputs. This learning approach requires labeling methods for establishing the *ground truth* in datasets and "learns" to identify patterns in the "known" training data. Typically, supervised learning solves *classification* and *regression* problems that pertain to predicting discrete (see Fig. 3.1) or continuous variables (see Fig. 3.2), respectively. For example, a regression problem involves predicting the delay in a data center network; algorithms that are practical for addressing it include Random Forest (RF) and Support Vector Machine (SVM). RF operates by constructing various decision trees during training, and a majority vote among the individual trees determines the final prediction; each tree is trained on a training data subset with a random subset of features. RF often outperforms individual decision trees, particularly when handling complex and high-dimensional datasets. SVM follows an algorithm that finds the optimal hyperplane that best separates data points from different classes, its decision function assigns new data points to one of the classes based on their position relative to such an hyperplane. In turn, a classification problem, for instance, is identifying mouse and elephant flows; a helpful technique to cope with this problem is the Adaptive Random Forest (ARF). ARF extends the traditional RF by incorporating concepts of adaptability and diversity to enhance its performance in streaming data scenarios.

Unsupervised Learning works with unlabeled training datasets and aims to create models that find patterns in the data. This kind of learning is most suited for addressing *clustering* problems (see Fig. 3.3). For instance, the K-means algorithm can be used to cluster large, long-lived data streams and small, short-lived data into elephant and mouse flows. K-means partitions a dataset into K distinct, non-overlapping subsets (or clusters) by minimizing the sum of the squared distances between centroids and data points.

Between Supervised and Unsupervised Learning resides ***Semi-supervised Learning*** helpful to operate with partial knowledge. A semi-supervised algorithm can be trained on a dataset that contains labeled and unlabeled samples (see Fig. 3.4); it is beneficial when obtaining labeled data is expensive or time-consuming and aims to leverage available labeled data to identify patterns, relationships, or structures and then generalize this knowledge to make predictions on the potentially abundant unlabeled data. Semi-Supervised Support Vector Machine (S3VM) extends SVM to find a decision boundary that not only separates labeled instances but also respects the inherent structure of the unlabeled data. SV3M can help detect anomalies in unmanned aerial vehicles where sensors provide less labeled data and more unlabeled data.

Self-Supervised Learning allows creating ML models from data without explicit external labeling, paving the path for training effective models with less reliance on labeled datasets. While semi-supervised learning combines labeled and unlabeled samples in a traditional supervised learning framework, self-supervised learning leverages the inherent structure or information within the input data to generate its supervision signals; essentially, the data provides supervision for training the model. By

Supplementary Information The online version contains supplementary material available at https://doi.org/10.1007/978-3-032-15176-6_3.

O. Caicedo Rendon, *Machine Learning for Network Management*, Textbooks in Telecommunication Engineering, https://doi.org/10.1007/978-3-032-15176-6_3

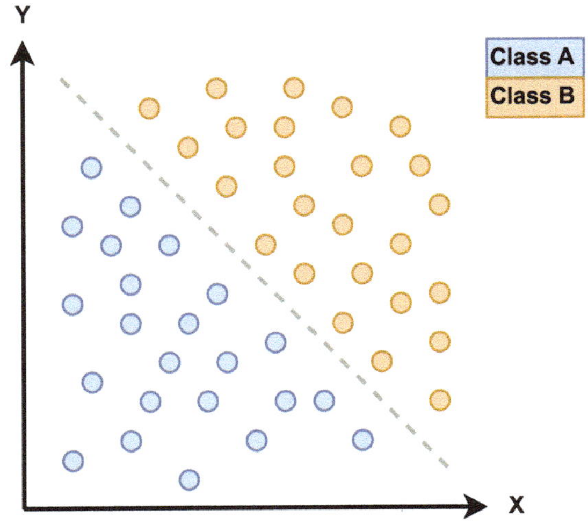

Fig. 3.1 Classification—predicting discrete outcomes

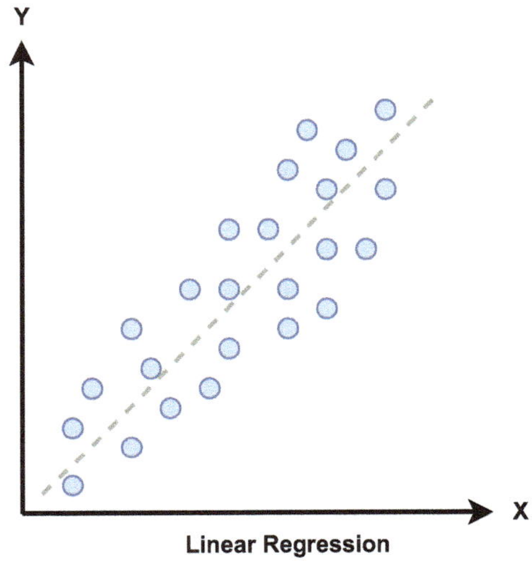

Fig. 3.2 Regression—predicting continuous outcomes

doing so, the model learns meaningful representations and features directly from the input data. Like semi-supervised learning, self-supervised learning is valuable when obtaining labeled data is challenging or expensive, such as when it is needed to classify encrypted network traffic in SDN.

Deep Learning techniques learn by discovering complex data structures and building their computational models using multiple processing layers; a Deep Neural Network (DNN) can create various levels of abstraction to represent data. Typical DL structures are Feedforward (FNN), Recurrent (RNN), and Graph (GNN) neural networks. FNN does not have cycles or loops in the neural network; the data moves from the input nodes through the hidden nodes and toward the output nodes. Convolutional Neural Networks (CNNs), the most well-known FNN model, has be applied for diverse network management tasks, including routing and traffic classification. An RNN is recursive in nature. The connections between neurons produce specific cycles, meaning that the output depends not only on their immediate inputs but also on the neuronal state of the previous step. An RNN is designed to use sequential data, i.e., it can relate the current state to the previous one. Long Short-Term Memory (LSTM) may be the most representative RNN [2] and can be used, for example, to predict network traffic. A GNN is specifically designed to operate on graph-based data, enabling prediction tasks at node, edge, and graph levels; it can be leveraged to perform efficient path selection in intent-driven networking. Advances in DL have enabled breakthroughs in areas like computer vision and robotics; in the networking domain, DL techniques are especially helpful in network resource management.

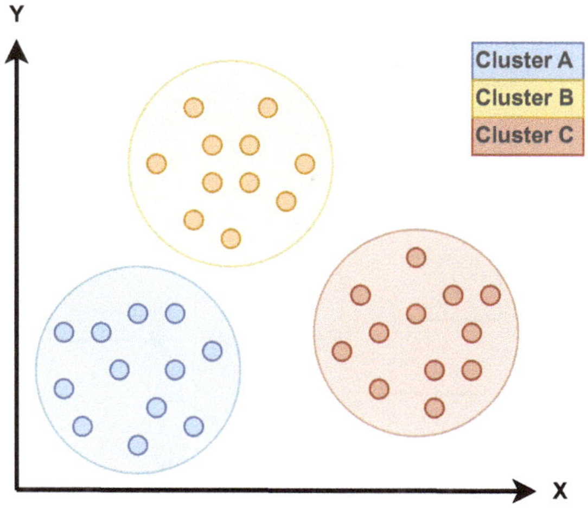

Fig. 3.3 Clustering—finding patterns

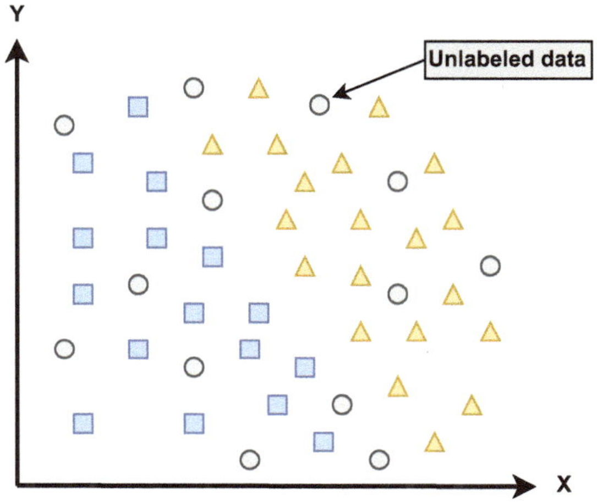

Fig. 3.4 Semi-supervised learning—unlabeled and labeled data

Reinforcement Learning, from a high-abstraction level, involves an agent that interacts with an environment (it represents the external world) to learn about it and exploit the learned knowledge (see Fig. 3.5a). The interactions are by actions that are rewarded or penalized. Therefore, the training data consists of a set of state-action pairs and rewards; an RL algorithm optimizes a cumulative reward by learning the policy (i.e., the best actions) through the feedback provided by the environment. Hence, this learning approach is best suited for making cognitive decisions. RL can be used, for example, to compute the optimal paths in an SDN by learning from past routing decisions to improve network performance in terms of delay, packet loss, and throughput. RL algorithms can require many interactions with the environment to achieve an optimal policy, generating an expensive computational process due to the amount of information stored and processed. As an RL extension, ***Deep Reinforcement Learning*** can cope with high-dimensional and extra-large action-state problems by using RL to train DL techniques (e.g., FNN and RNN) that timely learn optimal policies (see Fig. 3.5b). DRL algorithms can help enhance routing in large data center networks and resource management in 5G, 6G, and beyond networks.

Unlike discriminative learning, where the focus is on learning the decision boundary between different classes or categories in the data, ***Generative Learning*** primary objective is to capture the underlying distribution of the training dataset to generate realistic and novel samples that resemble its patterns and structures; it can take raw data (images, speech, and other complex data types) and learn to generate statistically probable outputs when prompted. Generative Adversarial Networks (GANs), Variational Autoencoders (VAEs), and Restricted Boltzmann Machines (RBMs) are generative learning techniques. A GAN comprises a generator that generates synthetic samples and a discriminator responsible for distinguishing between actual and generated samples; its training involves the two networks in a competitive setting, leading to the generation of increasingly

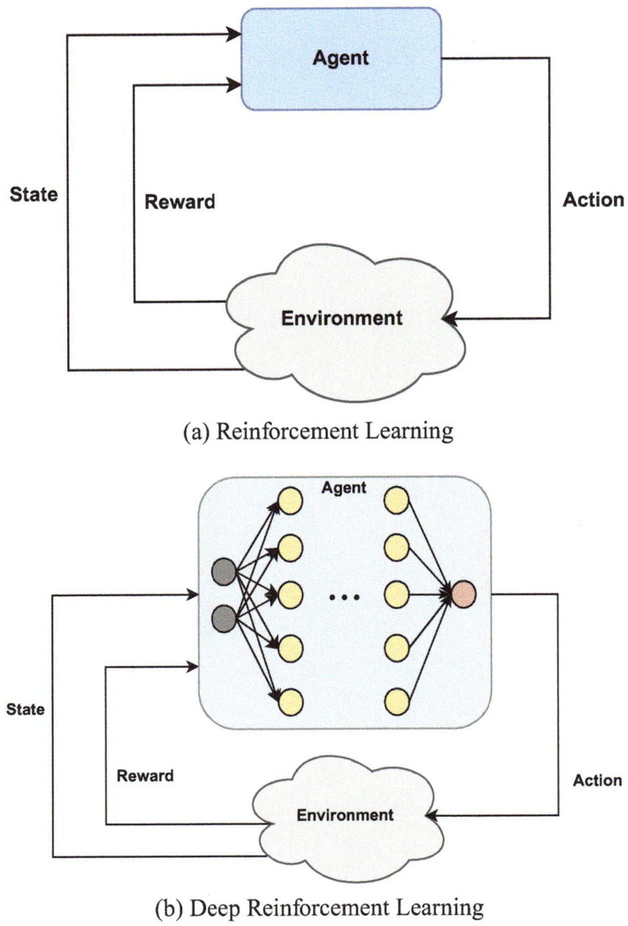

Fig. 3.5 Learning by experience

realistic samples. A VAE is an autoencoder that focuses on learning a probabilistic distribution from the input data; it encodes and decodes data while considering the inherent variability in the data distribution, enabling the generation of diverse samples. An RBM models the probability distribution of the input data to perform collaborative filtering and feature learning.

3.2 Evolution and Key Milestones

This section provides a brief history of ML, focusing on the techniques mainly applied to network management [3]. ML dawn dates back to 1943 when the first mathematical NN model for computers was proposed by McCulloch [4]. This model introduced the artificial neuron model, which is currently the core of NN development, requiring manual establishment of the weights of connections between neurons. Hebbian, in 1949, proposed an algorithm based on rules for updating weights in the initial NN model. The initial artificial neuron model and its automatic updating allowed the construction of the first NN computer in 1950 (Stochastic Neural Analog Reinforcement Computer). Alan Turing proposed, in the same year, a test to determine if a computer can achieve intelligent behavior [5].

The Ordinary Least Squares (OLS) was used in the 1950s to compute linear regressions in electromechanical desk calculators. In the same decade, two linear models for conducting classification were introduced: Maximum Entropy (MaxEnt) and Logistic Regression. Different ML techniques derived from pattern recognition and root-finding problems also appeared in the 1950s, exposing two non-parametric models for regression and classification: K-Nearest Neighbors (kNN) analyzes the data with a distance metric, and Kernel Density Estimation (KDE) estimates the probability density function of the data by applying a kernel function (e.g., Gaussian). Also in the same decade, research on root-finding optimization introduced the stochastic approximation algorithm that uses successive approximations to find the unique root of a regression function, later referred to as Stochastic Gradient Descent (SGD) since it approximates the proper gradient (computed using the whole dataset) by iteratively adding a scaled gradient estimate over every single instance of the dataset. SGD has become a well-known incremental (a.k.a., online) learning method for training various ML models, such as linear regression and NNs, facing large-scale datasets.

The Naïve Bayes (NB) algorithm (origins date back to the eighteenth and nineteenth centuries) applies Bayes' theorem on features with solid independence assumptions to perform probabilistic classification and was used in the 1950s for pattern recognition and information retrieval. NB was later generalized using KDE, also known as NBKE, to estimate the conditional probabilities of the features. In the area of clustering during the 1950s, the k-Means algorithm was also released; this algorithm uses a distance metric to partition the data into k subsets (clusters), where the distance to each cluster centroid is minimized. By the end of the 1950s, the Markov Decision Process (MDP) had been developed from the Markov model to formalize sequential decision-making problems in a fully observable, controlled environment, considering discrete-time states, transitions, actions, and rewards. Research efforts building on the initial NN model flourished too at that time: the modern concept of perceptron, a.k.a. FNN (the nodes from each layer connect directly to the nodes of the next layer), was introduced as the first NN model that could learn the weights from input samples. This model describes two NN classes according to the number of layers: Single Layer Perceptron (SLP) and Multi-Layer Perceptron (MLP). SLP is an NN with one input layer and one output layer. MLP is an NN with hidden layers (one or more) between the input and the output layers. In 1959, the term *"Machine Learning"* was coined by Arthur Samuel, who also developed a checkers-playing game recognized as the earliest self-learning program [6].

At the dawn of the 1960s, the Hidden Markov Model (HMM) algorithm emerged, operating in a partially observable, autonomous environment with conditional probabilities among hidden states and visible outputs. In the mid-1960s, the Baum-Welch algorithm was proposed to learn those conditional probabilities. An MDP extension called Partially Observable Markov Decision Process (POMDP) was proposed in 1965, and the algorithm (called Control Optimal POMDP), able to find optimal (or near-optimal) control strategies in partially observable stochastic environments, was proposed five years later. Finite Action-set Learning Automata (FALA) and Continuous Action-set Learning Automata RL algorithms (published in 1973) also extend MDP enhancement, and, according to the given rewards, continuously update the probabilities of making actions in an observed environment. The Automatic Interaction Detection (AID) algorithm was released in 1963 as the first DT to solve regression problems. AID divides an observation dataset sequentially into mutually exclusive subsets to reduce the error in predicting the target variable. Remarkably, applying DTs to classification problems started with the Theta Automatic Interaction Detection (THAID) algorithm a decade later. In 1965, the first algorithm for training MLP was published, marking the beginning of the DL era (the term started to be used in the 1980s in ML and during the 2000s, in the NNs domain). Minsky and Papert [7] carried out a mathematical analysis that highlighted the shortcomings of NNs based on perceptrons, which significantly reduced research interest in NNs by the end of the 1960s.

In the 1970s, various models surfaced, including Backpropagation, the Ebellar Model Arithmetic Computer (CMAC), Expectation-Maximization (EM), Temporal Difference, the Iterative Dichotomiser 3 (ID3), and AQ11. Up to now, BP has been the most popular NN training algorithm and comes in different flavors: SGD (the most used), conjugate gradient, one-step secant, Levenberg-Marquardt, and resilient BP. BP has a slow convergence speed and tends to fall into local optima; overall, its efficiency depends on the initial weights chosen. CMAC (released in 1975) has been extensively used in RL and classification problems because it learns faster than MLP. EM (proposed in 1977) generalizes the learning of the unknown parameters of statistical HMM models. TD (arose in 1977) is an RL algorithm that predicts the combination of immediate reward and its reward prediction at the next time; the changes in predictions drive the learning process. Around 1979, the second generation of DTs emerged with ID3 that performs attribute selection considering entropy[1] maximization; C4.5 and C5.0 algorithms (widely used in practice) are ID3's descendants, conceived to address shortcomings like data overfitting and training with missing values. AQ1 is a learning system that incrementally generates new rules using the existing ones and new training instances, which are later discarded after learning.

Thanks to a renewed interest in NNs, in the early 1980s emerged CNN, Self-Organized Maps (SOM), and the Hopfield network. CNN is an FNN envisioned originally for visual imagery analysis and classification; it requires minimal image preprocessing and connects neurons inspired by the organization of the animal visual cortex, where the visual field is divided between neurons, each responding to stimuli only in its corresponding region. Unlike traditional NNs that apply error-correction learning (such as SGD), SOM, designed for dimensionality reduction, follows an unsupervised competitive learning approach. The Hopfield network (introduced in 1982) is the first RNN; the weights connecting the neurons are bidirectional. Jordan in 1986 [8] offered the modern RNN's definition, an NN in which connections between neurons exhibit one or more

[1] Measure of the uncertainty about a source of messages.

cycles; the cycles provide a structure for internal states or memory, allowing RNNs to process arbitrary sequences of inputs. Therefore, RNNs are particularly useful in time series forecasting, handwriting recognition, and speech recognition.

After 1985, the BP algorithm's popularity increased notably thanks to its successful application in training DNNs and the emergence of novel NNs, such as RBM, Time-Lagged Feedforward Network (TLFN), and Radial Basis Function Neural Network (RBFNN). RBM is a Boltzmann machine variation in which there are no connections within any of the network layers (neurons form a bipartite graph), whether visible or hidden; this variation provides more efficient and straightforward learning than traditional Boltzmann machines. RBM is useful in various application domains, including dimensionality reduction, feature learning, and classification. The RBM's practicability increased significantly after the mid-2000s due to its ability to work with labeled and unlabeled datasets, as well as the introduction of Contrastive Divergence (a faster learning method for Boltzmann machines). TLFN is an MLP centered on time series forecasting. RBFNN is an NN with a weighted set of Radial Basis Function (RBF) kernels trainable in a supervised or unsupervised way.

Beyond advancements in NNs, several ML techniques thrived during the 1980s. Bayesian Networks (BN) arose as a Directed Acyclic Graph (DAG) representation for diverse statistical models (e.g., NB and HMM). The algorithm named Classification and Regression Trees (CART) was proposed to face classification and regression problems. The Reduced Error Pruning 3 (REPTree) algorithm was proposed to address classification problems with faster and simpler tree models based on information gain for splitting and reduced error pruning. DT also experienced its earliest incremental learning algorithms (e.g., ID4, ID5, and ID5R), which were built upon batch models (e.g., ID3), greatly influencing incremental DTs in the new millennium. Toward the end of the 1980s, two TD approaches were proposed for RL: TD (λ) and Q-learning. TD(λ) adds a discount factor ($0 \leq \lambda \leq 1$) that allows the policy evaluation process to determine to what extent estimates of previous state values, based on current errors, are eligible for updating. Q-learning replaces the traditional state-value function of TD with an action-value function (i.e., Q-value), which estimates the benefit of taking a specific action in a particular state. At the end of the decade, the application domains of ML started expanding to the operation and management of communication networks [9].

Significant advances were realized in NNs and DTs in the 1990s. GA and Particle Swarm Optimization (PSO), bio-inspired optimization algorithms, were used to train NNs. In this decade, one of the essential achievements in NNs was the work on LSTM, an RNN able to learn long-term dependencies for facing tasks involving long input sequences; LSTMs are particularly useful to perform speech recognition and Natural Language Processing (NLP). In 1992, the M5 and M5 Prime (M5P) algorithms were published for constructing tree-based multivariate linear models. The Least Absolute Shrinkage and Selection Operator (LASSO) regularization algorithm was also introduced around this time to enhance prediction accuracy in regression models.

SVM was also introduced in the 1990s to find an optimal solution in high-dimensional feature spaces; it enables the use of different kernel functions (e.g., linear, polynomial, and RBF). SVM is primarily used for classification; its regression variant is called Support Vector Regression (SVR). SVM-based classifiers find a hyperplane to discriminate between categories. A single-class SVM is a binary classifier; it deduces the hyperplane to differentiate the data belonging to the class from the rest of the data, that is, *one-vs-rest*. A multi-class approach in SVM can be formulated as various single-class classifiers. SVM can also learn incrementally using SGD, although this applies only to models that use a linear kernel function. By 1999, an incremental learning variant of SVM appeared for handling perceived and fundamental concept drifts.

Also, in the 1990s, the *ensemble learning* concept emerged to demonstrate that combining various single models can help enhance learning at the expense of computational overhead. The Weighted Majority Algorithm (WMA), Boosting, Adaptive Boosting (AdaBoost), Stochastic Gradient Boost (SGBoost), Bagging, and RF exemplify ensemble learning algorithms. RF focused on ensembling DT models. In RL, SARSA was introduced as a more realistic, however less practical, Q-learning variation. SARSA does not update the Q-value of an action considering the maximum action-value of the following state, as Q-learning does, but instead, it uses the Q-value of the action chosen in the next state. The advances given by the algorithms mentioned above encouraged the development of significant use cases in the 1990s, such as handwriting recognition and data mining. In this decade, Tom Mitchell defined ML as a computer program able to learn from experience [10].

The interest in SVM and ensemble DTs increased at the beginning of the twenty-first century, reflected in emerging algorithms such as the Multiplicative Algebraic Reconstruction Technique (MART), extra-trees, and eXtreme Gradient Boosting (XGBoost). MART and XGBoost follow Friedman's Gradient-Boosted Decision Trees (GBDT) algorithm, an ensemble DT algorithm based on gradient boosting. Extra-trees stands for *extremely randomized trees*; it is an ensemble DT algorithm that builds random trees based on k randomly chosen features and selects a split-point randomly for decreasing computational complexity. The popularity of DL increased after its application in the NN domain, but it declined due to the difficulty of training DNNs using BP and the growing competitiveness of techniques like SVM. Hinton's work on Deep Belief Networks (DBN) [11] gave a new breath and strength to research in DNNs in 2006. DBN introduced an efficient training strategy for DL, which was used successfully in different classes of DNN. The development of DNNs grew exponentially thanks to improvements in storage capacity and large-scale data processing. The wave of popularity in DL remains today; one approach that has recently received tremendous attention is DRL (a concept already discussed and applied 25 years ago), which incorporates DL models into RL for solving complex problems. In this regard, DQN (DNN plus Q-learning) was proposed for mastering video games [12].

The 2000s have also been fruitful for the incremental learning research community; several algorithms emerged, particularly for incremental DT. In 2001, the Very Fast Decision (VFDT) algorithm exposed an incremental DT that selects a root node from the first training instances and grows the tree from the leaf nodes. Concept-adapting VFDT (CVFDT) extended VFDT to address concept drift by maintaining a set of alternative DTs and storing instances over time, and, in subsequent years, incorporated NB classifiers in the tree's leaf nodes to improve accuracy. The Hoeffding Adaptive Tree (HAT) uses the Adaptive Windowing (ADWIN) algorithm to monitor the accuracy of the DT's branches, which are replaced with other, more accurate ones when their performance decreases. HOT introduced an ensemble method that utilizes additional option nodes to construct a single structure comprising multiple CVFDTs. The Adaptive HOT (AHOT) was designed as a HOT extension that stores error estimation in each leaf, computed using the Exponentially Weighted Moving Average (EWMA). ARF proposed another incremental ensemble DT, which adapts the widely used RF ensemble. In contrast to RF, ARF comes with a drift and warning detector per base DT, enabling selective resets and tree replacements when the warning becomes a drift.

In 2001, online bagging and online boosting were developed. Both incremental ensembles allow using different learners as the base model, such as NB and any incremental DT, though recent implementations use $\text{CVFDT}_{\text{NBC}}$. Around 2006, the online Passive–Aggressive (PA) algorithms were featured for solving classification and regression problems. These algorithms (*passively*) keep the same weights when no loss is present, but (*aggressively*) update them when the loss is positive. In 2011, two new approaches emerged: Very Fast Decision Rules (VFDR) and Accuracy Updated Ensemble (AUE). VFDR introduced an algorithm that, incrementally, learns ordered and unordered rule sets and classifies using the majority class strategy. A variant called VFDR_{NB} incorporated NB classifiers to improve accuracy. AUE presented an adaptive block-based ensemble for incremental learning, which selects and updates the base classifiers regarding the current distribution in a block series. Online AUE (OAUE), an AUE enhancement, introduced a new error-based weighting function to train and evaluate the base classifiers incrementally. The last implementations of both AUE and OAUE commonly use $\text{CVFDT}_{\text{NBC}}$ as the base learner.

The evolution of GAI spans several decades. In this path, Expert Systems (popularized in the 1960s and 1970s) can be considered among the earliest practical applications of generative AI. They used rule-based reasoning to replicate human expertise in specific domains and generate recommendations or solutions based on domain-specific knowledge. In the 1970s and 1980s, interest in NLP led to the development of systems that automatically generated reports, summaries, and even simple stories. In the 1980s and 1990s, the renewed interest in NNs and connectionism laid the foundation for modern ML techniques, and as a consequence, researchers explored NN-based approaches for generative tasks, such as image and speech synthesis. HMMs were widely used for generative tasks, in the 1980s and 1990s, such as speech synthesis. These models played a significant role in Text-to-Speech (TTS) systems, contributing to the improvement of computer-generated speech quality. In the 2000s, N-gram models [13] were employed for text generation tasks; they can rate text sequences based on the probability of each word or token following the previous ones, but they struggle to capture long-range dependencies and context. In the 2000s and 2010s, Probabilistic Graphical Models (PGMs), including BNs, became famous for generative tasks. PGMs represent complex probabilistic relationships in data to get a more accurate generation. The DL revolution ushered in GANs (introduced in 2014 in [14]), which gained prominence for their ability to generate, for instance, high-quality images and music, by training a generator network against a discriminator network. Transformers, a type of NN architecture introduced in 2017 in the paper "Attention Is All You Need" by Vaswani et al. [15], revolutionized NLP. In this regard, Large-scale Language Models (LLMs) [16] like GPT (Generative Pre-trained Transformer) have demonstrated exceptional text generation capabilities across an enormous range of applications, including mobile networks and network management configuration. In the 2020s and onward, GAI continues to evolve rapidly, with ongoing research focused on enhancing the quality and diversity of generated content while ensuring the responsible development of AI.

3.3 Methodologies for Machine-Learning Solutions

3.3.1 Building Supervised and Unsupervised Learning Solutions

Several approaches to building up ML-based solutions exist; some examples are the Knowledge Discovery in Database (KDD) [17], ML for networking pipeline [3], Cross-Industry Standard Process for Data Mining (CRISP-DM) [18], and CRISP-ML(Q) [19].

CRISP-ML(Q) is a methodology helpful for designing (semi) supervised and unsupervised ML-based solutions in the networking area; its main phases are:

- *Business, Problem, and Data Understanding* aim to ensure the feasibility of the ML project by identifying scope, verifying data quality, and defining success criteria. The criteria must be clear and measurable, they are typically defined as Key Performance Indicators (KPIs). The *Problem Understanding* involves a comprehensive grasp of the networking problem (e.g., flow clustering or delay prediction) that will be addressed by the knowledge generated by (semi-)supervised or unsupervised ML techniques. The *Data Understanding* involves collecting initial data (i.e., acquiring the necessary data), describing data (i.e., documenting the data), exploring data (e.g., identifying relationships), verifying data quality (e.g., documenting quality issues), defining the set of data and the set of classes required to address the outlined networking proble; it is essential since datasets vary from one problem to another and from one period to another.
- *Data Engineering* refers to building the final dataset. This phase primarily involves selecting data that refers to determining which data sets will be used and documenting reasons for features inclusion or exclusion, cleaning data (e.g., eliminating duplicate samples), constructing data (e.g., deriving new features that will be helpful), integrating data (e.g., creating data combining multiple data sources), formatting data (e.g., converting string values that store numbers to numeric ones to can perform mathematical operations), and selecting features. Feature selection is a crucial step that significantly impacts the computational cost and accuracy of estimating a target variable, such as network delay.
- *ML Model Engineering* is a journey of building one or multiple models (supervised or unsupervised algorithms) that meticulously analyze the complex relationships in data to predict one or more target variables like the throughput in a network slice. The process involves selecting modeling techniques, dividing the data into training, testing, and validation sets, constructing the models, and evaluating the candidate models to choose the one that satisfactorily performs the prediction. Notably, the candidates must be pitted against each other. The network practitioner and the pre-defined success criteria guide the interpretation of the model results, underscoring the importance of their role in the process.
- *ML Model Evaluation* aims to understand and analyze the results obtained in the previous steps, check the novelty and significance of the discovered knowledge, interpret the findings, and validate their impact. This phase includes evaluating outcomes in a test dataset for coping with various questions: (i) Do the models meet the business success criteria?, (ii) was anything overlooked?, and (iii) were all steps adequately executed? Here, it is also necessary to summarize the findings and correct errors as needed, determining whether to proceed to the Deployment step or iterate further.
- *Deployment* focuses on applying the discovered knowledge in an actual scenario by integrating the ML model into the existing network. This phase comprises planning deployment (documenting a deployment plan, including, for instance, hardware requirements for the inference task), planning monitoring and maintenance, producing a final report (involving, for example, user acceptance and usability), and reviewing the ML project (conducting a project retrospective about what went well, what could have been better, and how to improve).
- *Monitoring and Maintenance* enables active monitoring of the model's performance to detect degradation or malfunctioning with unseen data, and to determine whether re-training or modifications are necessary. Model's degradation can be influenced by the hardware and software stack used during development and deployment. During monitoring, it is necessary to analyze the business use case from a broad perspective, aiming to gain valuable insights for refining and aligning the ML solution more effectively with evolving business needs.

Batch Modeling

When using supervised and unsupervised ML approaches, data scientists primarily implement the classical *Batch Learning* approach: all data necessary to generate an inference is collected before training and simultaneously accessed [20]. Usually, *Batch Learning* divides data into *training*, *validation*, and *test* sets. The training dataset allows network practitioners to fit the parameters of an ML model (e.g., weights). With the validation dataset, network practitioners can choose suitable hyperparameters for an ML model (e.g., architecture, learning rate, regularization) or select a model from a pool of candidates. The test data set enables network practitioners to evaluate the unbiased performance of the chosen model. Remarkably, batch learning considers the data and its structure as static elements. A typical batch dataset division is $60/20/20\%$ for training, validation, and test data, or $70/30\%$ when validation is unnecessary. These rule-of-thumb divisions are reasonable for small datasets. Nevertheless, other divisions are feasible in datasets with millions of samples: $98/1/1\%$ or $99.8/0.1/0.1\%$.

In *Batch Learning*, validation and testing usually follow one of two methods (other evaluation methods for batch learning exist, such as leave-one-out and bootstrap): holdout or k-fold cross-validation. In the holdout method, a subset of the dataset is set aside for validation (or testing). In the k-fold cross-validation, network practitioners randomly divide the available dataset into k equal subsets. The validation (or testing) is repeated k times, with $k - 1$ unique subsets used for training and the remaining subset for validating (or testing) the model, and the outcomes are averaged over repetitions.

In its evolution, Batch learning has also incorporated upgrades, such as external storage and data subsets, that enable processing large but static datasets. However, the more data there is, the less performance to output a final static model. Moreover, batch learning fails to handle a continuous supply of changing data (i.e., sequential data or data stream), such as that generated by the high dynamism of networks. Since static models cannot continuously integrate new information, they must be re-constructed to ensure their validity over time. Remarkably, retraining a model from scratch is computationally expensive, time-consuming, and leads to potentially outdated models. An option to address batch learning shortcomings is *incremental* learning, where the model is retrained with only the new data.

Incremental Modeling

Continuously updating a model using sequential data (i.e., constantly arriving data with no specific order) without re-processing the data already used is called Incremental Learning [21].[2] Many datasets, although static, are so large that network practitioners would treat them as sequential data. As sequential data can become endless in specific domains, as networking, incremental learning seeks to balance model complexity and processing time. Such a balance enables lifelong learning and a constantly updated model within limited memory resources. Several fundamental aspects characterize an incremental learning algorithm: (i) it processes a data instance at a time in the arrival order, inspecting it as input only once,[3] (ii) it uses a limited amount of memory, even when processing data much more extensive than available memory, (iii) it restricts the runtime to a limit, which is particularly pivotal for algorithms aimed at real-time applications (e.g., networking); and (iv) it can provide a prediction anytime, no matter the number of instances used for training.

The evaluation of incremental learning algorithms can follow one of two methods [22]: the holdout method and the interleaved test-then-train method. Holdout represents a natural extension of batch learning, where a single, independent, and sufficiently large (i.e., tens of thousands of instances) test data set can provide a valid accuracy measurement. Network practitioners can evaluate the model on the test data set after the last training or periodically to track its performance. In scenarios where the data statistics change over time (i.e., concept drift), the test set must be dynamic; the data must be constantly collected using new instances that have not yet been used for training. Interleaved test-then-train (also known as prequential) refers to using each sample to test the model before training, and does not require a holdout set, allowing for the maximum utilization of the collected data for both testing and training. In interleaved test-then-train, early mistakes from a poorly trained model (also known as a cold start) can negatively impact the actual accuracy that incremental algorithms might achieve. Although this punishment diminishes over time, evaluation techniques like pre-training and sliding windows help correct the accuracy measurement. Network practitioners usually use interleaved test-then-train with incremental learning algorithms, as it easily handles a potentially infinite sequence of instances arriving one after another.

---

[2] The definition of incremental learning is not always consistent in the literature and involves certain ambiguity regarding related terms, including online learning, data stream mining, stream learning, and incremental online learning. The definition presented in this book covers the commonalities among such related concepts.

[3] An algorithm may store some examples internally in the short term. However, at some point, some of them need to be discarded to meet memory and time limits.

3.3.2 Explaining Learning Models

Nowadays, most networking solutions based on ML employ opaque models, as they are more proficient in capturing nonlinear patterns and solving complex tasks than white-box models that are inherently explainable. The opacity produces models that humans cannot easily understand, considerably hindering the model's adoption in real deployments. The Explainable Artificial Intelligence (XAI) aims to promote the adoption of ML-based solutions in real-life, including networking-focused ones, by providing interpretability, explainability, and transparency to opaque models [23]. Interpretability is the degree to which a human being can understand the response of a system. Explainability refers to the ability to get, within the domain being interpreted, features that have contributed to a sample's production of a decision; it implies generating representations for these interpretations. Transparency refers to the system's ability to provide explanations that humans can understand within the deployment context; note that transparency encompasses both explainability and interpretability.

XAI solutions can be categorized based on their scope (i.e., Local or Global) and the method of explanation extraction (i.e., Intrinsic or Post hoc). *Local* embraces solutions targeted to explain a specific decision or a set of predictions. The Local Interpretable Model-agnostic Explanations (LIME) exeplifie this kind of solutions and focuses on training local surrogate models to approximate and explain individual predictions performed by a black-box model. *Global* comprises solutions that explain the model logic as a whole. The Model Understanding through Subspace Explanations (MUSE) solution facilitates understanding any black-box model by describing how it behaves in subspaces characterized by features of interest. *Intrinsic* XAI includes models, such as Naïve Bayes and DTs, inherently interpretable, simple, and self-explanatory by design, and can be readily presented to humans through simulation, decomposition, or algorithmic analysis. *Post hoc* XAI embraces solutions focused on analyzing the black-box model after training by creating, for instance, a simpler white-box model that can explain the original one.

Visualization-oriented, simplification-oriented, and Feature-relevance-oriented categories are also helpful for classifying explainable post hoc ML-based solutions. The most straightforward XAI post hoc solution involves explaining the model's internal operations and interactions through visualizations, which entail visual augmentation and (optionally) dimension reduction to generate illustrations. Visualizations can be used to create graphs to show the inference process and highlight the most salient features of a DL model performing traffic classification. The model simplification XAI solutions build a functionally similar model, but one that is much simplified (e.g., DT-based models), to elucidate the inference process. Simplification-oriented solutions can interpret decisions based on DL performed in video streaming, flow scheduling, and routing solutions in SDN. Feature relevance analysis XAI solutions compute a feature relevance score to assess how each feature impacts the final decision made by opaque models. The Shapley Additive exPlanations (SHAP), a solution based on cooperative game theory, gives scores to features involved in a learning model and can be used to interpret an RL model that computes paths in a large SDN.

Explainable RL (XRL) and Explainable DRL (XDRL) have garnered particular attention in diverse domains, including robotics and games. Achieving XRL and XDRL for the Computer Networks and Telco Industry is pivotal to deploying RL and DRL-based solutions in real networks; nonetheless, applying XAI to provide transparency, interpretability, and explainability to the decision-making process performed by RL and DRL agents is an open research challenge. Remarkably, explainability is challenging to achieve in RL and DRL because agents learn autonomously without human intervention by interacting with the environment to develop policies or values. In this regard, it is necessary to explain the policy learned (i.e., why an action was performed); this is complex since high-complexity RL and DRL algorithms are commonly parameterized with thousands, if not millions, of parameters. Additionally, XAI could aid in addressing the current limitations of these algorithms to accelerate the development of new ones. Considering the part of the learning agent under explanation, XRL and XDRL solutions can be categorized into *Feature Importance*, *Learning Process and MDP*, and *Policy Level*. *Feature Importance* comprises solutions designed to identify the features that influence an agent's choice of action in a specific environment's state. *Learning Process and MDP* presents solutions that demonstrate influential experiences or the components of the MDP that significantly impact how the agent behaves. *Policy Level* refers to solutions intended to illustrate how an agent behaves in the long term.

Network practitioners usually follow the methodologies described in Sect. 3.3.1 to build explainable models intrinsically. To apply post hoc XAI (including interpretability, credibility, and effectiveness) to black-box models, the practitioners need to carry out additional phases to those defined by the methodologies mentioned above, namely Get Explanations and Model Improvement (see Fig. 3.6); the TRUSTEE methodology introduced those phases [24].

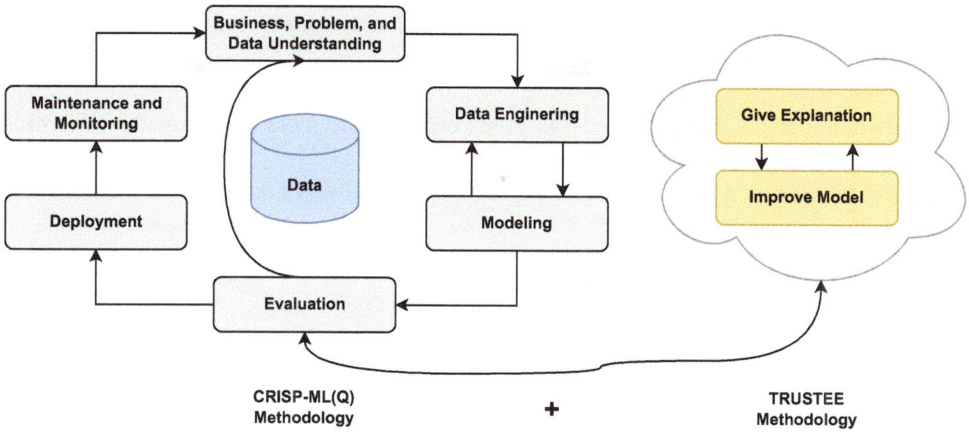

Fig. 3.6 XAI methodology

- *Get Explanation.* The network practitioner must choose and enforce at least one of the XAI methods mentioned in this Section. The TRUSTEE framework (developed with the application domain of network security in mind) can be used in this phase, for instance, to extract a DT-based and high-fidelity white-box learning model from any black-box learning model.
- *Model Improvement.* The network practitioner must assess the performance of the explanations obtained in the previous step to determine whether the black-box learning model evaluated is credible and trustworthy. Remarkably, this last step relies on the knowledge of networking experts and represents an open issue in the networking domain. Among other aspects, it is necessary to evaluate the accuracy and fidelity of the explanations. Furthermore, the results obtained by XAI enforcement can be leveraged to optimize the opaque ML models, for instance, by eliminating features or samples according to Shapley values or LIME scores.

3.4 Summary

This chapter provided initially a journey through the main ML paradigms, which are powerful tools that can be used to solve clustering, regression, classification, and decision-making in network management. Then it guided through the fascinating evolution and key milestones of ML, a testament to the rapid progress in this field. Finally, this chapter equipped you with methodologies to create and explain supervised, unsupervised, semi-supervised, and self-supervised learning solutions. Notably, using well-established and novel ML techniques can lead to groundbreaking advancements in autonomous network management tasks, aiming to achieve zero-touch network management and self-driving networks.

Conceptual Questions

3.1 Define what machine learning is.

3.2 Explain how each machine-learning paradigm operates.

3.3 Discuss how batch and incremental learning operates.

3.4 Explain the path from AI to Generative AI.

3.5 What is explainable artificial intelligence? Why is it relevant for realizing autonomous network management tasks?

3.6 What methodological steps should be followed to build supervised and unsupervised machine-learning solutions? Also, detail the tasks to perform in each step.

3.7 Explain what a white-box machine learning model is. Also, explain what a black-box machine-learning model is?

3.8 What methodological steps should be followed to explain supervised and unsupervised machine-learning solutions? Also, detail the tasks to perform in each step.

Solutions of This Chapter

3.1 Problem. Define what machine learning is.

Solution. ML is a knowledge area that focuses on developing algorithms to enable computers to perform tasks without explicit instructions. ML algorithms learn patterns and make data driven decisions.

3.2 Problem. Explain how each machine-learning paradigm operates.

Solution. Supervised learning uses labeled training data sets to create models that map inputs to their corresponding outputs. This learning approach requires labeling methods for establishing the ground truth in data sets and learning to identify patterns in the training data. Unsupervised learning works with unlabeled training datasets to create models that find patterns in the data. Semi-supervised learning operates with partial knowledge. It is beneficial when obtaining labeled data is expensive or time-consuming and aims to leverage available labeled data to identify patterns, relationships, or structures and then generalize this knowledge to make predictions on the potentially abundant unlabeled data. Self-supervised learning includes ML algorithms that allow models to learn from the data without explicit external labeling; this type of learning offers a promising avenue for training effective models with less reliance on labeled datasets. While semi-supervised learning combines labeled and unlabeled samples in a traditional supervised learning framework, self-supervised learning leverages the inherent structure or information within the input data to generate its supervision signals; essentially, the data provides supervision for training the model. By doing so, the model learns meaningful representations and features directly from the input data.

3.3 Problem. Discuss how batch and incremental learning operates.

Solution. All data necessary to generate an inference is collected before training and simultaneously accessed in batch learning. Usually, this type of learning divides data into training, validation, and test sets. The training dataset allows network practitioners to fit the parameters of an ML model (e.g., weights). With the validation dataset, the network practitioners can choose the suitable hyperparameters of an ML model (e.g., architecture, learning rate, regularization) or select a model from a pool of candidates. The test data set allows network practitioners to assess the unbiased performance of the model chosen. Remarkably, batch learning considers the data and its structure as static elements. In turn, continuously updating a model using sequential data (i.e., constantly arriving data with no specific order) without re-processing the data already used is called incremental learning. The incremental learning models usually are created by interleaved test-then-train. Interleaved test-then-train refers to using each sample to test the model before training. Note that this method does not need a holdout set, making maximum use of the collected data for testing and training. In interleaved test-then-train, early mistakes from a poorly trained model (a.k.a., cold start) punish the actual accuracy that incremental algorithms might achieve. Although this punishment diminishes over time, evaluation techniques like pre-training and sliding windows help correct the accuracy measurement.

3.4 Problem. Explain the path from AI to GAI.

Solution. AI-based systems aim to simulate human intelligence in machines programmed to perform tasks requiring reasoning, learning, perception, and decision-making. Generative AI-based systems, an extension of AI-based ones, refer to systems that create new content, including text, audio, video, images, music, and programming code. This content creation is enabled by generative models that can learn the dataset distribution and create new samples. The generative model's appearance was a natural progression in AI, driven by advances in DL, computational power, and the availability of large datasets in diverse knowledge areas.

3.5 Problem. What is explainable artificial intelligence? Why is it relevant for realizing autonomous network management tasks?

Solution. Most networking solutions based on AI operate opaque models since they are more proficient in capturing nonlinear patterns and solving complex tasks. The opacity considerably hinders the model's adoption in autonomous network management and, overall, communication and telecommunication networks. XAI aims to face opacity and provide interpretability, explainability, and transparency to AI models. Interpretability is the degree to which a network practitioner can understand the response of an AI-based networking system. Explainability refers to the offer of visual representations of interpretations. Transparency is the system's ability to generate explanations that network practitioners understand. From the characteristics described above, it is possible to perceive that transparency incorporates explainability and interpretability.

3.6 Problem. What methodological steps should be followed to build supervised and unsupervised machine-learning solutions? Also, detail the tasks to perform in each step.

Solution. Constructing supervised, self-supervised, unsupervised, and semi-supervised ML-based systems involve various phases: Business, Problem, and Data Understanding, Data Engineering, ML Model Engineering, ML Model Evaluation, Deployment, and Monitoring and Maintenance. Business, Problem, and Data Understanding aim to ensure the feasibility of the ML project by identifying its scope, data quality verification, and success criteria; these criteria must be clear and measurable, usually defined as Key Performance Indicators (KPIs). Business and problem understanding involve a comprehensive grasp of the networking problem with the knowledge generated by ML. Data Understanding comprises creating, gathering, and defining the set of data and classes needed to cope with the outlined networking problem; it is essential since datasets vary from one problem to another and from one period to another. Data Understanding involves collecting raw data, describing data, exploring data relationships, and verifying data quality. Data Engineering refers to building the final dataset. This phase primarily consists of selecting data, cleaning data, constructing data, integrating data, formatting data, and selecting features. Feature selection is a crucial step that significantly impacts the computational cost and accuracy of estimating a target variable in any area, including network management. ML Model Engineering is a journey of building one or multiple models that meticulously analyze the complex relationships in data to predict one or more target variables. Modeling involves selecting candidate ML models, dividing the data into training, testing, and validation sets, constructing the models, and evaluating the candidates to choose the most optimally performing the prediction. In such a selection process, multiple models are pitted against each other, and it is the networking practitioner's domain knowledge, with the pre-defined success criteria, guides the interpretation of the model results, underscoring the importance of their role in the process. ML Model Evaluation aims to understand the previous step's results, check the novelty and significance of the discovered knowledge, interpret the results, and validate their impact. This phase includes evaluating outcomes in a test dataset. The deployment phase focuses on applying the knowledge discovered in an actual scenario by integrating the ML model into the network management system. This phase comprises planning, monitoring, and executing the deployment, producing a final report, and reviewing the ML project. Monitoring and maintenance is the pivotal phase after deploying the ML model into production to monitor its performance, detect its degradation actively or malfunctioning with unseen data, and determine whether re-training or modifications are necessary. The degradation can also be influenced by hardware and software stacks. In addition to monitoring and re-training, reflecting on the business use case and the specific ML task is beneficial. This reflection can provide valuable insights for refining the ML-based solution and its construction methodology to align better with evolving business needs.

3.7 Problem. Explain what a white-box machine learning model is. Also, explain what a black-box machine learning model is?

Solution. A white-box ML model (e.g., linear regression and decision trees) is a model where the internal workings are entirely understandable and explainable; they are designed to provide clear insights into how predictions or decisions are made. In this regard, they are interpretable and transparent. Interpretability means network practitioners can easily understand the model's logic, parameters, and decision-making process. Transparency means the relationships between input features and predictions can be described visually (preferably), facilitating the understanding of network practitioners. Conversely, to a white-box ML model, a black-box ML model's (e.g., DNNs, RandomForest, and SVM) internal workings are not easily interpretable or transparent; the relationships between input features and predictions are not explicitly defined or easily visualized. In sum, these models are often highly complex and rely on sophisticated algorithms that make it difficult to understand how predictions or decisions are made. The choice between white-box and black-box ML depends on the specific task. White-box models can be preferred when interpretability, transparency, and regulatory compliance are critical. On the other hand, if the primary goal is to achieve high accuracy on complex tasks and interpretability is less relevant, black-box models can be the best option.

3.8 Problem. What methodological steps should be followed to explain supervised and unsupervised machine-learning solutions? Also, detail the tasks to perform in each step.

Solution. Network practitioners should follow two steps to explain and improve black-box (opaque) ML models: Get Explanation and Model Improvement. In Get Explanation, the network practitioner must choose and apply XAI methods, such as LIME, SHAP, PDP, and surrogate models. In Model Improvement, the network practitioner must assess the performance of the explanations obtained in the previous step. Determining whether the black-box learning model evaluated is credible and trustworthy is also essential. Remarkably, this last step depends on networking experts' knowledge and represents an open issue in the networking domain. Among other aspects, evaluating the explanations' accuracy and fidelity is necessary. Furthermore, the results obtained by XAI enforcement can be leveraged to optimize the opaque ML models, for instance, by eliminating non-relevant features or samples according to Shapley values or LIME scores.

References

1. S. Russell, P. Norvig, *Artificial Intelligence: A Modern Approach*, 3rd edn. (Prentice Hall Press, Upper Saddle River, 2009)
2. F.A. Gers, J.A. Schmidhuber, F.A. Cummins, Learning to forget: continual prediction with LSTM. Neural Comput. **12**(10), 2451–2471 (2000)
3. R. Boutaba, M.A. Salahuddin, N. Limam, S. Ayoubi, N. Shahriar, F. Estrada-Solano, O.M. Caicedo, A comprehensive survey on machine learning for networking: evolution, applications and research opportunities. J. Internet Serv. Appl. **9**(1), 16 (2018)
4. W.S. McCulloch, W. Pitts, A logical calculus of the ideas immanent in nervous activity. Bull. Math. Biophys. **5**(4), 115–133 (1943)
5. A.M. Turing, Computing machinery and intelligence. Mind **59**(236), 433–460 (1950)
6. A.L. Samuel, Some studies in machine learning using the game of checkers. IBM J. Res. Dev. **3**(3), 210–229 (1959)
7. M.L. Minsky, S. Papert, *Perceptrons: An Introduction to Computational Geometry* (MIT Press, Cambridge, 1972)
8. M.I. Jordan, Serial order: a parallel distributed processing approach, in *Technical Report ICS Report 8604* (University of California, San Diego, 1986)
9. A. Jennings, A learning system for communications network configuration. Eng. Appl. Artif. Intell. **1**(3), 151–160 (1988)
10. M.T. Mitchell, *Machine Learning*, 1st edn. (McGraw-Hill Inc, New York, 1997)
11. G.E. Hinton, S. Osindero, Y.W. Teh, A fast learning algorithm for deep belief nets. Neural Comput. **18**(7), 1527–1554 (2006)
12. V. Mnih, K. Kavukcuoglu, D. Silver, A.A. Rusu, J. Veness, M.G. Bellemare, A. Graves, M. Riedmiller, A.K. Fidjeland, G. Ostrovski, S. Petersen, C. Beattie, A. Sadik, I. Antonoglou, H. King, D. Kumaran, D. Wierstra, S. Legg, D. Hassabis, Human-level control through deep reinforcement learning. Nature **518**(7540), 529–533 (2015)
13. E.M. De Novais, T.D. Tadeu, I. Paraboni, Improved text generation using n-gram statistics, in *Advances in Artificial Intelligence-IBERAMIA 2010: 12th Ibero-American Conference on AI, Bahía Blanca, Argentina, November 1–5, 2010. Proceedings 12* (Springer, New York, 2010), pp. 316–325
14. I. Goodfellow, J. Pouget-Abadie, M. Mirza, B. Xu, D. Warde-Farley, S. Ozair, A. Courville, Y. Bengio, Generative adversarial nets. Adv. Neural Inform. Process. Syst. **27**, 1254 (2014)
15. A. Vaswani, N. Shazeer, N. Parmar, J. Uszkoreit, L. Jones, A.N. Gomez, Ł Kaiser, I. Polosukhin, Attention is all you need. Adv. Neural Inform. Process. Syst. **30**, 5426 (2017)
16. T. Mikolov, A. Deoras, D. Povey, L. Burget, J. Černocký, Strategies for training large scale neural network language models, in *Proceedings of the 2011 IEEE Workshop on Automatic Speech Recognition and Understanding* (IEEE, 2011), pp. 196–201
17. U. Fayyad, G. Piatetsky-Shapiro, P. Smyth, From data mining to knowledge discovery in databases. AI Mag. **17**(3), 37 (1996). https://doi.org/10.1609/aimag.v17i3.1230
18. P. Chapman, J. Clinton, R. Kerber, T. Khabaza, T. Reinartz, C. Shearer, R. Wirth, *Crisp-dm 1.0 Step-by-Step Data Mining Guide* (SPSS inc, 2000)
19. S. Studer, T.B. Bui, C. Drescher, A. Hanuschkin, L. Winkler, S. Peters, K.R.M. Iler, Towards crisp-ml(q): a machine learning process model with quality assurance methodology. Mach. Learn. Knowl. Extract. **3**(2), 392–413 (2021)
20. M. Bramer, *Principles of Data Mining*, 2nd edn. (Springer, Berlin, 2013)
21. A.R.T. Gepperth, B. Hammer, Incremental learning algorithms and applications, in ESANN (2016)
22. A. Bifet, G. Holmes, R. Kirkby, B. Pfahringer, Data Stream Mining A Practical Approach (The University of Waikato, 2019), p. 5
23. G.A. Vouros, Explainable deep reinforcement learning: state of the art and challenges. ACM Comput. Surv. **12**, 1652 (2022)
24. A.S. Jacobs, R. Beltiukov, W. Willinger, R.A. Ferreira, A. Gupta, L.Z. Granville, AI/ML for network security: the emperor has no clothes, in *Proceedings of the 2022 ACM SIGSAC Conference on Computer and Communications Security, CCS '22* (Association for Computing Machinery, New York, NY, 2022), pp. 1537–1551. https://doi.org/10.1145/3548606.3560609

Part II
Network Management Driven by Learning Models

The book's second part presents the main aspects of network management driven by learning models. In this way, Chap. 4 reviews how supervised learning assists network management. Chapter 5 discusses network management assisted by unsupervised learning. Chapter 6 presents deep reinforcement learning for network management. Chapter 7 introduces the uses of generative learning for network management.

4 Network Management Assisted by Supervised Learning

4.1 Linear Models

Linear models, such as simple, multiple, polynomial, Lasso, and Ridge, are fundamental tools in regression analysis. Network practitioners can use these models to solve various network management tasks. Some of them are (i) predicting bandwidth usage trends, helping network administrators plan for capacity upgrades, and ensuring optimal performance. (ii) Foreseeing throughput for coping with capacity planning and performance optimization by considering features as network traffic and protocol types. (iii) Forecasting traffic loads for different network segments to enhance load balancing and resource allocation. (iv) Estimating QoS violations by identifying links or network segments where performance metrics violate quality thresholds. (v) Predicting bandwidth requirements while avoiding overfitting to ensure the model generalizes well to new data.

4.1.1 Linear Regression

Linear regression is an ML algorithm widely used for predicting a continuous dependent variable (target), such as the delay in a network link and the network's throughput. When this algorithm considers one independent variable (feature), it is referred to as a simple linear regressor. If the algorithm involves two or more features, it is referred to as a multiple linear regressor. These types of regression assume a linear relationship between the feature(s) and the target, as shown in Eq. 4.1. Where y is the target. $x_i, \ldots, x_n$ are features. β_0 is the intercept and represents the expected value of y when all x_i are zero. $\beta_1, \beta_2, \ldots, \beta_n$ are the slopes for the respective features; each β_i represents the change in y for a one-unit change in x_i when the other variables are hold constant. The slopes (β) are estimated by minimizing the Eq. 4.2 that computes the Ordinary Least Squares (OLS) cost function, i.e., the summation of the squared differences between actual values y_i and predicted values $\hat{y}_i$ during m independent observations.

$$y = \beta_0 + \beta_1 x_1 + \beta_2 x_2 + \cdots + \beta_n x_n \tag{4.1}$$

$$J(\beta) = \sum_{i=1}^{m} (y_i - \hat{y}_i)^2 = \sum_{i=1}^{m} (y_i - (\beta_0 + \beta_1 x_{i1} + \cdots + \beta_n x_{\text{in}}))^2 . \tag{4.2}$$

4.1.2 Polynomial Regression

Polynomial regression is used to make predictions when the features in a dataset exhibit a curvilinear relationship with the target. This relationship is modeled via Eq. 4.3 that represents an n-th degree polynomial. Here, $\beta_0, \beta_1, \beta_2, \ldots,$ and β_n are coefficients obtained, for instance, with OLS, ϵ is the error that assesses the variability that the model cannot explain about the independent variable y, and n defines the flexibility of the model. It is noteworthy that a high-order polynomial often

Supplementary Information The online version contains supplementary material available at https://doi.org/10.1007/978-3-032-15176-6_4.

O. Caicedo Rendon, *Machine Learning for Network Management*, Textbooks in Telecommunication Engineering, https://doi.org/10.1007/978-3-032-15176-6_4

leads to overfitting and high computational costs. Furthermore, the polynomial complexity can hinder the interpretation of coefficients.

$$y = \beta_0 + \beta_1 x + \beta_2 x^2 + \cdots + \beta_n x^n + \epsilon. \tag{4.3}$$

4.1.3 Regularization Techniques

Ridge and Lasso are regularization techniques used to improve the model's generalization to new data by addressing overfitting by preventing the coefficients from becoming too large. Ridge Regression (or L2 regularization) does not perform feature selection because it shrinks coefficients rather than eliminates them, and is given by Eq. 4.4. Where y_i and $\hat{y}_i$ are the actual and predicted values, m and n are the number of observations and features, β_j are the coefficients, and λ is the regularization parameter. Remarkably, the term ($\lambda \sum_{j=1}^{n} \beta_j^2$) discourages the use of high coefficients, as it increases the penalty for high coefficients; thus, small coefficients are preferred.

$$J(\beta) = \sum_{i=1}^{m} (y_i - \hat{y}_i)^2 + \lambda \sum_{j=1}^{n} \beta_j^2. \tag{4.4}$$

Lasso Regression aims to prevent overfitting and is given by Eq. 4.5, where y_i and $\hat{y}_i$ are the actual and predicted values, m and n are the number of observations and features, respectively, β_j are coefficients, and λ is the parameter used for regularization. This technique discourages high coefficients, as it adds the absolute values of them to the loss function, leading to a preference for small coefficients. Lasso is suitable when the dataset is expected to include a large number of features, but only a subset of them is essential for prediction, as it carries out feature selection. In contrast, Ridge is preferable when all features in the dataset are expected to contribute to forecasting.

$$J(\beta) = \sum_{i=1}^{m} (y_i - \hat{y}_i)^2 + \lambda \sum_{j=1}^{n} |\beta_j|. \tag{4.5}$$

4.1.4 Latency Prediction

Simple linear regression for network latency prediction

In network performance management, with historical data, a simple linear regression model can relate the quantity of active users (independent feature, X) to network latency (target, y): $y = \beta_0 + \beta x$. Listing 4.1 presents a snippet of Python code to address the raised example.The model developed enables network administrators to predict how latency increases with the number of active users in the network, thereby facilitating proactive management and resource allocation.Note that the exemplified Python code can serve as a template for the reader to address other linear regression problems by modifying the lines associated with dataset creation.Second,as the objective of the example is not to obtain the best model,the evaluation metrics are explained at the end of this section.

```
#Import necessary libraries
import numpy as np
import pandas as pd
from sklearn.model_selection import train_test_split
from sklearn.linear_model import LinearRegression
from sklearn.metrics import mean_squared_error, r2_score
import matplotlib.pyplot as plt

#Assume a dataset 'network_data' with 'active_users' and 'network_latency
    ' columns
data = {
    'active_users': [10, 20, 30, 40, 50, 60, 70, 80, 90, 100],
    'network_latency': [5, 7, 8, 12, 15, 17, 19, 22, 24, 27]
}
network_data = pd.DataFrame(data)

```

```
#Define the independent feature (X) and target (y)
X = network_data[['active_users']]
y = network_data['network_latency']

#Split the data into training and testing sets
X_train, X_test, y_train, y_test = train_test_split(X, y, test_size=0.3, random_state=42)

#Create and train the simple linear regression model
model = LinearRegression()
model.fit(X_train, y_train)

#Make predictions on the testing set
y_pred = model.predict(X_test)

#Evaluate the model
mse = mean_squared_error(y_test, y_pred)
r2 = r2_score(y_test, y_pred)
print(f"Mean Squared Error: {mse:.2f}")
print(f"R-squared: {r2:.2f}")

#Visualize the results
plt.scatter(X, y, color='blue', label='Actual data')
plt.plot(X_test, y_pred, color='red', linewidth=2, label='Regression line')
plt.xlabel('Number of Active Users')
plt.ylabel('Network Latency (ms)')
plt.title('Simple Linear Regression: Network Latency vs Active Users')
plt.legend()
plt.show()
```

Listing 4.1 Linear Regressor

4.1.5 Throughput Prediction

Multiple linear regression for network throughput prediction

In network performance management, multiple linear regression can predict network throughput (y, the rate at which data is successfully transmitted from one point to another in a network over a given period) based on several factors (i.e., features). For instance, network throughput can be influenced by the number of active users x_1, the type of application (x_2, e.g., video streaming), and the time of day (x_3). With historical data and by defining $y = \beta_0 + \beta_1 x_1 + \beta_2 x_2 + \beta_3 x_3$, the model would allow network practitioners to understand the combined impact of different independent factors (features) on network throughput and, as a result, make informed decisions to optimize performance. Listing 4.2 presents a snippet of Python code to address the raised example. The reader can use this code as a template to develop others multiple linear regressors by essentially oodifying the lines associated with the dataset creation.

```
#Import necessary libraries
import numpy as np
import pandas as pd
from sklearn.model_selection import train_test_split
from sklearn.linear_model import LinearRegression
from sklearn.metrics import mean_squared_error, r2_score
import matplotlib.pyplot as plt

#Load the data
#Assume 'network_data' with 'active_users', 'application_type', 'time_of_day', and 'network_throughput' columns
data = {
    'active_users': [10, 20, 30, 40, 50, 60, 70, 80, 90, 100],
    'application_type': [1, 0, 1, 0, 1, 0, 1, 0, 1, 0],  # 1 for video streaming, 0 for other types
    'time_of_day': [9, 12, 15, 18, 21, 0, 3, 6, 9, 12],  # time in hours
```

```
    'network_throughput': [100, 150, 200, 250, 300, 350, 400, 450, 500,
    550]
}
network_data = pd.DataFrame(data)

#Define the independent features (X) and target (y)
X = network_data[['active_users', 'application_type', 'time_of_day']]
y = network_data['network_throughput']

#Split the data into training and testing sets
X_train, X_test, y_train, y_test = train_test_split(X, y, test_size=0.3,
    random_state=42)

#Create and train the multiple linear regression model
model = LinearRegression()
model.fit(X_train, y_train)

#Make predictions on the testing set
y_pred = model.predict(X_test)

#Evaluate the model
mse = mean_squared_error(y_test, y_pred)
r2 = r2_score(y_test, y_pred)
print(f"Mean Squared Error: {mse:.2f}")
print(f"R-squared: {r2:.2f}")

#Visualize the results
plt.figure(figsize=(10, 6))
plt.scatter(y_test, y_pred, color='blue', edgecolor='k', alpha=0.7, label
    ='Predicted vs Actual')
plt.plot([y.min(), y.max()], [y.min(), y.max()], 'k--', lw=2, label='
    Ideal fit')
plt.xlabel('Actual Network Throughput')
plt.ylabel('Predicted Network Throughput')
plt.title('Actual vs Predicted Network Throughput')
plt.legend()
plt.show()

# Plot residuals
plt.figure(figsize=(10, 6))
residuals = y_test - y_pred
plt.scatter(y_pred, residuals, color='red', edgecolor='k', alpha=0.7)
plt.axhline(0, color='k', linestyle='--', lw=2)
plt.xlabel('Predicted Network Throughput')
plt.ylabel('Residuals')
plt.title('Residuals of Predictions')
plt.show()
```

Listing 4.2 Mutiple Linear Regressor

4.1.6 Network Traffic Prediction

Polynomial regression and Lasso regularization for network traffic prediction

In network performance management, polynomial regression can be combined with Lasso regularization to model complex, curvilinear relationships in network traffic (the volume of data moving across a network at any given time) patterns. In this regard, network traffic (y) may exhibit daily peaks (x, time of day) and troughs that cannot be accurately obtained by a linear model: $y = \beta_0 + \beta_1 x + \beta_2 x^2 + \beta_3 x^3 + \epsilon$. A model based on this equation can capture the cyclical nature of network traffic, helping administrators predict and manage traffic loads more effectively. Polynomial regression is particularly useful for identifying trends and patterns that vary over time. Listing 4.3 shows a snippet of Python code that shows how to use polynomial regression with Lasso regularization to solve the raised example. The code is in accordance with the methodology defined in Chap. 3 and creates a sample dataset, defines the the time as the independent feature (X) and the network traffic as the target (y), and normalizes the features with StandardScaler that follows standard normal distribution (it makes $mean = 0$ and scales the data to unit variance). Furthermore, the code creates polynomial features, splits the dataset into training and testing sets, creates a Lasso regression model that is trained using the polynomial features, uses the built model to make predictions on the testing set, evaluates the model's performance using MSE and R^2, prints such metrics, and visualizes the actual data and the predicted trend. As mentioned earlier, the exemplified code can be used by readers as a template to face polynomial regression problems.

```
#Import necessary libraries
import numpy as np
import pandas as pd
from sklearn.model_selection import train_test_split
from sklearn.preprocessing import StandardScaler, PolynomialFeatures
from sklearn.linear_model import Lasso
from sklearn.metrics import mean_squared_error, r2_score
import matplotlib.pyplot as plt

#Load and preprocess the data
#Assume a dataset 'network_data' with 'time_of_day' and 'network_traffic' columns
data = {
    'time_of_day': [0, 1, 2, 3, 4, 5, 6, 7, 8, 9, 10, 11, 12, 13, 14, 15, 16, 17, 18, 19, 20, 21, 22, 23],
    'network_traffic': [10, 12, 15, 18, 25, 30, 35, 50, 55, 60, 70, 80, 85, 90, 80, 70, 60, 50, 40, 30, 25, 20, 15, 12]
}
network_data = pd.DataFrame(data)

# Define the independent feature (X) and target (y)
X = network_data[['time_of_day']]
y = network_data['network_traffic']

#Normalize the features
scaler = StandardScaler()
X_scaled = scaler.fit_transform(X)

#Create polynomial features
poly = PolynomialFeatures(degree=3)
X_poly = poly.fit_transform(X_scaled)

#Split the data into training and testing sets
X_train, X_test, y_train, y_test = train_test_split(X_poly, y, test_size=0.3, random_state=42)

#Polynomial Regression with Lasso Model
lasso_model = Lasso(alpha=0.1)
lasso_model.fit(X_train, y_train)
y_pred = lasso_model.predict(X_test)

#Evaluate the model
mse = mean_squared_error(y_test, y_pred)
r2 = r2_score(y_test, y_pred)
print(f"Polynomial Regression with Lasso - Mean Squared Error: {mse:.2f}")
print(f"Polynomial Regression with Lasso - R-squared: {r2:.2f}")

#Visualize the results
plt.scatter(X['time_of_day'], y, color='blue', label='Actual data')
plt.plot(X['time_of_day'], lasso_model.predict(poly.transform(scaler.transform(X))), color='red', label='Predicted trend')
plt.xlabel('Time of Day')
plt.ylabel('Network Traffic')
plt.title('Polynomial Regression with Lasso: Network Traffic Prediction')
plt.legend()
plt.show()
```

Listing 4.3 Mutiple Linear Regressor

Evaluation metrics for regression models

R-squared (R^2—see the corresponding Python code in Listing 4.1, Line 31) measures the proportion of variance in the target that is predictable from the features, helping to assess the performance of ML-based regression, including linear, multiple, and polynomial models. R^2 is given by Eq. 4.6, where the sum of the squared errors ($\text{SS}_{\text{res}} = \sum_{i=1}^{m}(y_i - \hat{y}_i)^2$) measures the variability of the target variable non explained by the model and $\text{SS}_{\text{tot}} = \sum_{i=1}^{m}(y_i - \bar{y})^2$ quantifies the variation in the actual data around the mean of the target ($\bar{y}$). When R^2 equals 1, the features explain all the variability of the target, while if it equals 0, the features used by the model do not explain any variability. Furthermore, an R^2 value close to 1 indicates a good fit between the model and the data.

$$R^2 = 1 - \frac{\text{SS}_{\text{res}}}{\text{SS}_{\text{tot}}}. \tag{4.6}$$

As high R^2 values can hide overfitting (e.g., adding more features to the model can inflate R^2 without necessarily improving the model) and sometimes be misleading, the performance evaluation of a regression model must involve other metrics, including Mean-Squared Error (MSE), Root MSE (RMSE), and Mean Absolute Error (MAE). MSE measures the average of the squared differences of the predicted ($\hat{y}_i$) and actual values (y_i) and is given by Eq. 4.7 where m is the number of observations. Note that the squaring of errors ensures that more significant errors have a disproportionate impact on the MSE, making it sensitive to outliers and less interpretable than RMSE and MAE, because MSE is expressed in the squared units of the target variable. A low MSE indicates predictions are close to the actual values. A high MSE indicates a poor fit, i.e., predictions deviate significantly from the actual values.

$$\text{MSE} = \frac{1}{m}\sum_{i=1}^{m}(y_i - \hat{y}_i)^2. \tag{4.7}$$

The RMSE metric measures the average magnitude of the errors between the predicted values and the actual values; see Eq. 4.8 where y_i are the actual values, $\hat{y}_i$ are the predicted ones, and m is the number of observations. This metric is handy because it provides error information in the same units as the target variable; therefore, an RMSE around 0 indicates a pretty good fit of the model to the data. RMSE is sensitive to outliers (it squares the differences between actual and predicted values, which can disproportionately affect the metric) and does not indicate whether the model tends to over-predict or under-predict.

$$\text{RMSE} = \sqrt{\frac{1}{m}\sum_{i=1}^{m}(y_i - \hat{y}_i)^2}. \tag{4.8}$$

Other metric widely used to measure the performance of regression models is MAE. This metric is calculated by Eq. 4.9, where m is the number of observations. MAE provides information about the average magnitude of the errors i.e., the absolute difference between the actual (y_i) and predicted ($\hat{y}_i$) target values. In this regard, an MAE near 0 can indicate high performance of the model. Remarkably, MAE is less sensitive to outliers than RMSE because it does not square the errors, does not provide information on the variance of the errors, and does not indicate whether the predictions are biased.

$$\text{MAE} = \frac{1}{m}\sum_{i=1}^{m}|y_i - \hat{y}_i|. \tag{4.9}$$

4.2 Support Vector Machines

SVMs enable network practitioners to address diverse network management challenges, including predicting bandwidth and delay. The prediction of bandwidth available in a particular link or the network can aid in planning activities for capacity upgrades and ensuring optimal network performance. The forecasting of the delay enables the creation of network policies to improve the user experience of delay-aware applications. Other problems addressable with SVM include (i) modeling throughput from features as active users and protocol types, predicting throughput aids in capacity planning and performance optimization, (ii) forecasting traffic loads for different network segments, such as Edge and Core, to enhance load balancing and resource allocation, (iii) classifying network traffic according to application classes (e.g., video, voice, gaming, and browsing) to support services prioritization and, overall, QoS management, (iv) identifying user profiles (e.g., residential and business) based on activity patterns, enabling personalized services and targeted marketing; and (v) detecting QoS violations by identifying links or segments where network performance metrics unmet quality thresholds.

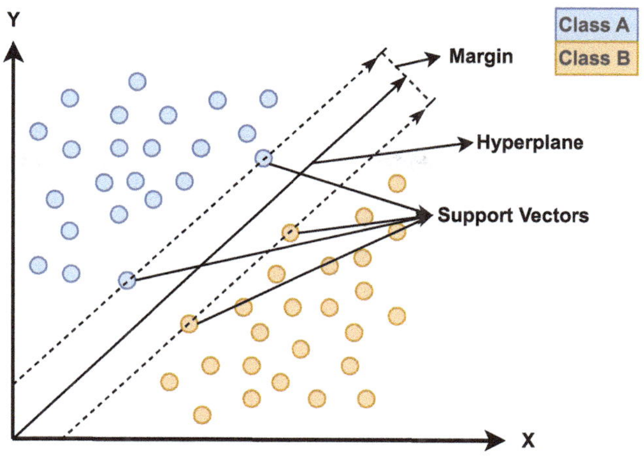

Fig. 4.1 Nonlinear support vector machine

4.2.1 Nonlinear Support Vector Machine

SVM (see Fig. 4.1) is a supervised robust ML algorithm for classification and regression tasks. This algorithm is based on the hyperplane concept, which generalizes the concept of a plane to higher-dimensional spaces. If the work implies a d-dimensional space, a hyperplane is a $(d-1)$—dimensional subspace. To exemplify, in a 2-dimensional space (a plane), a hyperplane is a 1-dimensional line that divides the plane into two regions. For a classification problem, each region corresponds to a class. Equation 4.10) represents mathematically the raised example, where $w_1, w_2, \ldots, w_d$ are the coefficients (weights) of the hyperplane that determines the orientation of the line, $x_1, x_2, \ldots, x_d$ are the coordinates of points in the term, and b is the bias term that shifts the line.

$$w_1x_1 + w_2x_2 + \cdots + w_dx_d + b = 0. \tag{4.10}$$

In classification tasks, the hyperplane is the decision boundary that separates the classes; therefore, the SVM algorithm needs to find the optimal hyperplane that maximizes the margin between them. The margin measures the distance between the hyperplane and the support vectors (i.e., the nearest data points for each class). For datasets without linear relationships, SVM finds a hyperplane (a line in 2D, a plane in 3D) that divides the classes. For nonlinear data, SVM, using a kernel function (e.g., RBF), maps the input features into a higher-dimensional space. The kernel trick allows this transformation without explicitly computing the coordinates in the higher-dimensional space. In regression tasks, instead of finding a single hyperplane, SVM finds a function that has at most a specific deviation (ϵ-tube) around the hyperplane within which the predictions of continuous values are considered acceptable. Note that data points outside this tube contribute to the error, which SVM aims to minimize. Furthermore, support vectors are the data points that lie on the boundary of the ϵ-tube or outside it; these points define the position and orientation of the hyperplane.

Algorithm 1 Non-linear SVM Classifier

Input: Dataset D with features X and labels y
Output: SVM model M
Procedure:
Select a kernel function K (e.g., RBF, Polynomial)
Transform the dataset D into a higher-dimensional space using K
Find the optimal hyperplane that maximizes the margin between classes in the transformed space
for each support vector **do**
    Compute the decision boundary based on the support vectors
end for
End Procedure

The nonlinear SVM classifier (see Algorithm 1) includes:

- The dataset, which is transformed by a kernel function such as RBF.
  - RBF is a kernel that transforms the original feature space into an infinite-dimensional space. SVM uses the transformation performed by RBF to create a linear decision boundary in a higher-dimensional space, corresponding to a nonlinear boundary in the original space. This kernel function can be calculated using Eq. 4.11, where x_i and x_j represent the input feature vectors, $\|x_i - x_j\|$ denotes the Euclidean distance between these vectors, and γ represents the influence of a training sample. When the γ value is low, the influence is far, resulting in smoother decision boundaries. A high value of γ indicates a substantial impact, resulting in more complex and tightly fitted decision boundaries.
  $$K(x_i, x_j) = \exp\left(-\gamma \|x_i - x_j\|^2\right). \tag{4.11}$$
  - It is noteworthy that RBF does not require explicit feature transformation, making it computationally efficient. Furthermore, this kernel can handle nonlinear relationships between features, and provides flexibility by adjusting the parameter γ. Nevertheless, choosing the right γ is crucial and challenging; remarkably, RBF can lead to overfitting if γ is too high.
- The algorithm finds the optimal hyperplane that maximizes the margin between classes in the transformed space.
- The decision boundary is computed on the basis of the support vectors, and it separates the different classes in the transformed space.
- Nonlinear SVM can be highly effective for complex datasets where the relationship between features and target variables is not linearly separable. However, this algorithm requires careful tuning of hyperparameters, such as the choice of kernel and γ, and can be computationally intensive for large datasets.

4.2.2 Anomaly Detector

SVM for anomaly detection

In network security management, detecting anomalies in network traffic is a common task that helps identify potential security threats or network failures. Nonlinear SVM can classify network traffic data into normal and anomalous traffic. According to the methodology discussed in Chap. 3, it is necessary (i) to collect network traffic data, including features such as packet size, flow duration, protocol type, number of bytes sent, and number of bytes received. (ii) Labeling the data as "normal" or "anomalous" by using historical analysis or expert knowledge. (iii) Selecting relevant features for detecting network anomalies and normalizing or standardizing the features to ensure they are on a similar scale. (iv) Splitting the dataset into training and testing. (v) Selecting an appropriate kernel function. (vi) Training the SVM Model to learn to find the optimal hyperplane that separates regular traffic from the anomalous one. (vii) Evaluating the performance of the trained SVM model with metrics such as accuracy and precision. (viii) Once the model is trained and evaluated, it can be deployed to monitor live network traffic and detect anomalies in real time. Listing 4.4 shows a Python code snippet to realize the above-described traffic classification with a nonlinear SVM. This code can be used by readers as a template to face other classification problems.

```
#Import necessary libraries
import numpy as np
import pandas as pd
from sklearn.svm import SVC
from sklearn.model_selection import train_test_split
from sklearn.preprocessing import StandardScaler
from sklearn.metrics import classification_report, accuracy_score,
    precision_score, f1_score

#Create a sample dataset
```

```
data = {
    'packet_size': np.random.randint(50, 1500, 300),
    'flow_duration': np.random.normal(5, 2, 300),
    'protocol_type': np.random.choice(['TCP', 'UDP', 'ICMP'], 300),
    'bytes_sent': np.random.randint(100, 10000, 300),
    'bytes_received': np.random.randint(100, 10000, 300),
    'label': np.random.choice(['normal', 'anomalous'], 300)
}

network_data = pd.DataFrame(data)

#Encode categorical data
network_data['protocol_type'] = network_data['protocol_type'].map({'TCP':
    0, 'UDP': 1, 'ICMP': 2})

#Define features and labels
features = network_data[['packet_size', 'flow_duration', 'protocol_type',
    'bytes_sent', 'bytes_received']]
labels = network_data['label']

#Normalize the features
scaler = StandardScaler()
features_scaled = scaler.fit_transform(features)

#Split the data into training and testing sets
X_train, X_test, y_train, y_test = train_test_split(features_scaled,
    labels, test_size=0.3, random_state=42)

#Choose a kernel function (RBF kernel)
svm_model = SVC(kernel='rbf', gamma='scale')

#Train the SVM model
svm_model.fit(X_train, y_train)

#Evaluate the model
y_pred = svm_model.predict(X_test)

#Compute accuracy, precision, and F1-score
accuracy = accuracy_score(y_test, y_pred)
precision = precision_score(y_test, y_pred, pos_label='anomalous')
f1 = f1_score(y_test, y_pred, pos_label='anomalous')

#Print evaluation metrics
print(f"Accuracy: {accuracy:.2f}")
print(f"Precision: {precision:.2f}")
print(f"F1-score: {f1:.2f}")
print(classification_report(y_test, y_pred))

#Deploy the model for real-time anomaly detection, it would involve
    integrating the trained model into the network monitoring system
```

Listing 4.4 SVM Classifier

Evaluation metrics for classification models

Accuracy is a learning performance metric that measures the percentage of correctly classified instances out of the total number of cases. In turn, the Precision metric assesses the model's ability to predict anomalies correctly. i.e., the correct outcomes. Accuracy and Precision are given by Eqs. 4.12 and 4.13). In these equations, framed within the example developed in the above code, True Positives (TPs) are cases in which the model correctly predicts an anomaly when there is a one actually exists. True Negatives (TNs) refer to cases where the model correctly predicts the absence of an anomaly when there is no anomaly in the traffic. False Positives (FPs) means the model predicts an anomaly, but there is no actual anomaly in the network traffic. False Negatives (FNs) are cases where the model fails to detect an anomaly in the network traffic.

$$\text{Acurracy} = \frac{\text{TP} + \text{TN}}{\text{TP} + \text{TN} + \text{FP} + \text{FN}} \tag{4.12}$$

$$\text{Precision} = \frac{\text{TP}}{\text{TP} + \text{FP}} \tag{4.13}$$

4.3 Decision Trees and Ensembles

Decision Tree and Ensemble Tree algorithms support similar tasks to those aided by SVM, such as bandwidth utilization prediction, network latency forecasting, network throughput forecasting, traffic load estimation, normal and malicious traffic classification, service traffic identification, user behavior detection, and connected device identification. Another classical task supported by these algorithms is flow classification into heavy hitters and non-heavy hitters. According to the total number of bytes transmitted in the flow, heavy hitters can be classified as elephants or mice; this classification is pivotal for handling network congestion and improving routing protocols. Simple DTs are helpful for initial models and provide easy-to-interpret results. They can handle regression and classification tasks but may suffer from overfitting and be less accurate for complex tasks than algorithms, such as RF, XGBoost, and AdaBoost, that combine (ensemble) multiple decision trees to improve accuracy, robustness, and generalization.

4.3.1 Decision Tree

DTs (see Fig. 4.2) are of the utmost importance in supervised ML algorithms. A DT can be used for classification and regression tasks. Typical classification and regression tasks in network management addressed by DTs include detecting anomalous and regular traffic, as well as predicting network latency in wireless networks. The DT algorithm models decisions and their possible consequences as a tree-like structure. Such a structure comprises a root node, internal nodes, branches, and leaf nodes. The root node represents the dataset, which is split into subsets. Internal nodes represent features used to separate the data. Branches represent the outcome decision that leads to further nodes. Leaf nodes represent the final class labels (decision outcomes).

Algorithm 2 DT Classifier

Input: Dataset D with features X and labels y
Output: Decision tree T
Procedure:
Select the best feature A to split the data using a criterion (e.g., Gini impurity, Information Gain)
Split the dataset D into subsets D_v based on the selected feature A
for each subset D_v **do**
  if stopping condition is met **then**
    Assign a class label to the leaf node based on the majority class of the instances in D_v
  else
    Recursively apply the procedure to D_v
  end if
end for
End Procedure

The DT classifier (see Algorithm 2) includes:

- The feature that best splits the dataset is selected based on a criterion such as Gini impurity.
  - A classical DT splits the data using the feature that minimizes the Gini impurity of the resulting child nodes. The Gini impurity, computed by Eq. 4.14, seeks to ensure that the nodes created after the split are as pure as possible.

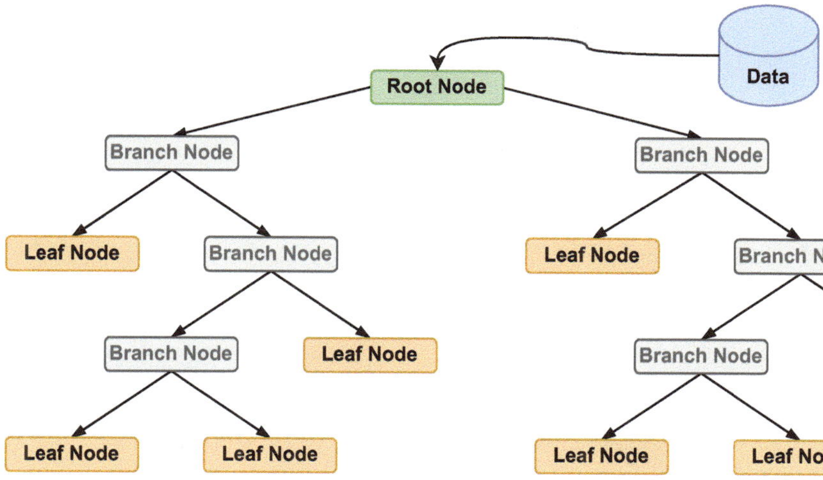

Fig. 4.2 Decision tree

Here, D is the data at the node. Furthermore, C is the total number of classes and $p_i = \frac{n_i}{N}$ is the proportion of the number of instances of class n_i concerning the total number of samples in D. A low Gini value indicates a high degree of purity—a node is pure if all the samples belong to the same class and, as a result, the impurity is 0.

$$\text{Gini}(D) = 1 - \sum_{i=1}^{C} p_i^2. \tag{4.14}$$

– Notably, the Gini impurity is straightforward to calculate and easy to interpret, works well for binary classification problems, and tends to create more significant and more homogeneous splits (balanced trees). Despite the advantages offered by Gini, it has relevant limitations related to sensitivity to class imbalance (suboptimal trees for minority classes), disregarding the probability distribution of the classes (less informative splits), generating overfitting in complex datasets unless pruning techniques are applied, and difficulties in operating in scenarios where different classes have different misclassification costs.

- The dataset is divided into subsets based on the selected feature. Note that splits aims to minimize the variance within the subsets.
- Recursively apply the above steps to each subset until one of the stopping conditions is met. Examples of such conditions are achieving the maximum tree depth, achieving the maximum number of nodes, attaining the minimum number of samples per leaf node, and reaching a pure node.
- The algorithm assigns a class label to each leaf node by voting (considering the majority class of the instances in that node). The process is similar for regression tasks, but the leaf nodes contain numerical values instead of class labels. The final prediction for a given input is got by averaging the values at the leaf nodes corresponding to the input's path through the tree.
- Remarkably, a DT requires few computational resources, can handle numerical and categorical data, and does not make assumptions about the data distribution. Furthermore, as DT is a white-box ML algorithm, it is easy to understand and interpret. Despite its advantages, a DT is susceptible to suffering from overfitting with deep trees. Additionally, it can be unstable, meaning that small changes in the data can lead to different trees.

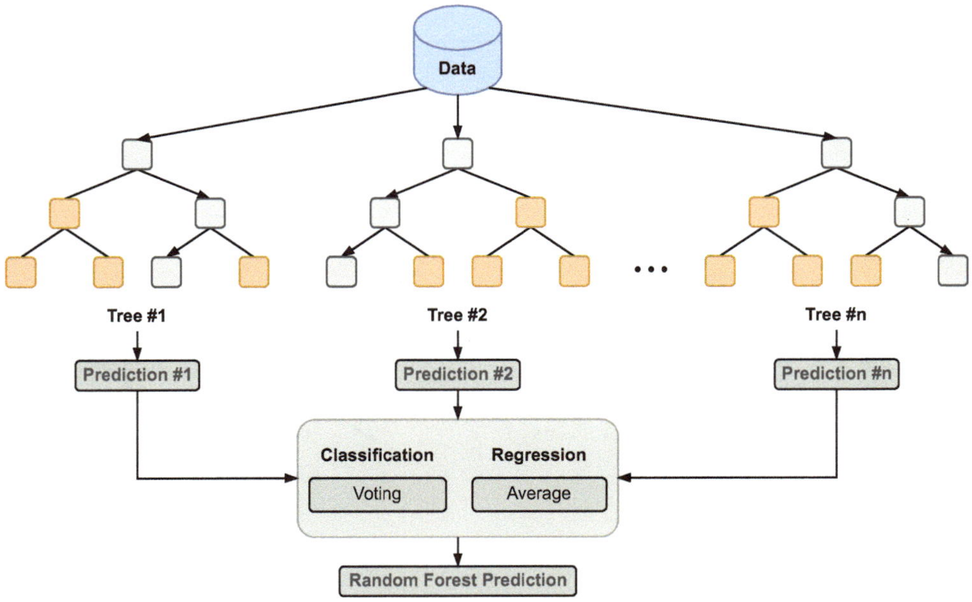

Fig. 4.3 Random forest

4.3.2 Random Forest

RF (see Fig. 4.3) is a highly versatile and robust ensemble supervised ML algorithm helpful in carrying out classification and regression tasks. This algorithm combines multiple decision trees to improve accuracy and control overfitting. On the one hand, when using RF for classification tasks, such as elephant flow detection, each decision tree in the forest provides a class prediction, and the final output is the class that is most frequently predicted by all individual trees (i.e., voting). On the other hand, when using RF for regression tasks, such as traffic size prediction, each decision tree in the forest provides a numerical prediction, and the final output corresponds to the average of the forecasts from all individual trees.

Algorithm 3 Random Forest Classifier

Input: Dataset D with features X and labels y, number of trees N
Output: Random forest model F
Procedure:
for $i = 1$ to N **do**
    Draw a bootstrap sample D_i from the dataset D
    Initialize the root node with D_i
    Grow a decision tree T_i on D_i recursively, $T_i \leftarrow$ GROW_TREE(root, 0)
    Add T_i to the forest F
end for
Aggregate the predictions of the N trees to form the final prediction
Define the recursive function GROW_TREE:
function GROW_TREE(node, depth)
    if stopping condition is met **then**
        Assign a class label to the leaf node based on the majority class of the instances in *node*
        return *node*
    else
        Randomly select a feature A
        Randomly select a split value v between the minimum and maximum values of A
        Split the data at *node* into *left* and *right* subsets based on v
        $node.left \leftarrow$ GROW_TREE($left, depth + 1$)
        $node.right \leftarrow$ GROW_TREE($right, depth + 1$)
        return *node*
    end if
end function
End Procedure

The RF classifier (see Algorithm 3) includes:

- The feature that best splits the dataset is selected based on a criterion such as Information Gain from a random subset of features.
  - The RF algorithm splits the dataset with the feature that maximizes the Information Gain of the resulting child nodes. This gain aims to ensure that the nodes created are as informative as possible. Information Gain is given by Eq. 4.15, where D represents the data at the node, A is the attribute under consideration, Values(A) are the possible values of A, D_v is the subset of D for which attribute A has value v, and Entropy(D) is the entropy of D. Recall, that Entropy measures the uncertainty or randomness in a probability distribution. For a discrete random variable X with possible outcomes $x_1, x_2, \ldots, x_n$ and a probability function $P(X)$, the entropy $H(X)$ is given by Eq. 4.16.

$$\text{Gain}(D, A) = \text{Entropy}(D) - \sum_{v \in \text{Values}(A)} \frac{|D_v|}{|D|} \text{Entropy}(D_v) \tag{4.15}$$

$$H(X) = -\sum_{i=1}^{n} P(x_i) \log P(x_i) \tag{4.16}$$

  - In summary, Information Gain considers the probability distribution of the classes for determining the optimal split in a decision tree, resulting in more informative splits. Despite its advantages, Information Gain presents shortcomings, as it can be computationally intensive for large datasets, and it may not be very objective towards features with many distinct values or categories.
- The dataset is divided into subsets based on the selected feature.
- Recursively apply the above steps to each subset until one of the stopping conditions is met. Examples of stopping conditions include, on the one hand, maximum_depth_reached to prevent the tree from becoming too deep and overfitting the training data, and, on the other hand, minimum_samples_node to ensure that nodes do not have too few samples, which can also lead to overfitting.
- Assign a class label to each leaf node based on the majority class of the instances in that node.
- Aggregate the predictions of all decision trees to generate the final prediction by using voting or averaging. Remarkably, RF is an ensemble method that uses multiple trees to produce more accurate and robust predictions; the aggregation helps reduce overfitting and improves the model's accuracy. Furthermore, it can effectively handle large datasets with higher dimensionality. Despite the advantages offered by RF, it can be computationally intensive and require significant resources during training when considering a large number of trees and features.

4.3.3 eXtreme Gradient Boosting

XGBoost is an advanced gradient boosting ML algorithm valuable for classification and regression tasks—this algorithm creates groups of DTs sequentially. In the sequence, each tree tries to correct the errors of the previous one. For classification tasks, such as predicting streaming or web traffic, XGBoost outputs the class with the highest probability of occurrence. In turn, it outputs the sum of the predictions from all trees for regression tasks, as forecasting the end-to-end delay in a remote surgery service in a 6G network. XGBoost is known for its efficiency, scalability, and high performance.

Algorithm 4 XGBoost Classifier

Input: Dataset D with features X and labels y, number of boosting rounds N
Output: XGBoost model F
Procedure:
Initialize the model with a base prediction
for $i = 1$ to N **do**
  Compute the residuals (errors) for the current model
  Fit a decision tree T_i to the residuals
  Update the model by adding the predictions of T_i weighted by a learning rate
end for
End Procedure

The XGBoost classifier (see Algorithm 4) includes:

- Initialize the model with a base prediction, which can often be the mean of the target variable for regression tasks or the log odds for binary classification.
- Iteratively improve the model by fitting new decision trees to the residual errors of the current model.
  - Each tree is fit to minimize the residuals (the differences between the actual and predicted values), which is crucial to ensure that each new tree focuses on the areas where the model is currently performing poorly.
  - The learning rate controls the updates to the model. It scales the contribution of each new tree, seeking to prevent overfitting and model generalization when new data is presented.
- The final prediction is the sum of the base prediction and the weighted contributions of all the trees.
- Advanced techniques incorporated by XGBoost include regularization to mitigate overfitting, parallelization for efficient processing, and pruning to reduce complexity. This tree algorithm can also handle missing values internally, making it robust to incomplete datasets. The techniques mentioned above make XGBoost a powerful and flexible ML algorithm capable of handling large datasets with high-dimensional features. Remarkably, it is challenging to set up XGBoost for achieving optimal learning performance due to its numerous hyperparameters. Note that finding the optimal combination of parameters can be time-consuming and requires extensive experimentation and domain expertise, resulting in high computational resource consumption. Another challenge when using XGBoost is that models created by this algorithm can be complex to interpret; the combination of many trees makes the overall model complex and less transparent. Furthermore, despite regularization techniques (Lasso and Ridge), XGBoost can still overfit when the number of boosting rounds (trees) is too high.

4.3.4 Classifying Streaming Traffic

DT, RF, and XGBoost for Classifying Streaming Traffic

DT, RF, and XGBoost can classify streaming application traffic in network performance management, helping network administrators optimize resource allocation and meet SLAs. Following the methodology discussed in Chap. 3 and beyond, the collection, data pre-processing, and feature engineering steps are necessary to (i) create a dataset with features that include the flow duration, total number of forwarding packets, total backward of packets, total length of forwarding packets, total length of backward packets, maximum length of forwarding packets, minimum length of forwarding packets, and type of transport protocol. As suggested by its name, flow duration feature represents the total time of the flow from the first packet to the last packet. The total number of forwarding packets represents the total number of packets sent from the source (in the forward direction) during the flow, providing insight into the source's volume of data; this feature is essential for understanding communication patterns. The total backward of packets counts the packets sent from the destination (in the backward direction) to the source during the flow; it helps in analyzing the response behavior of the destination and identifying asymmetric communication patterns. The total length of forwarding packets is the total length (in bytes) of all packets sent in the forward direction, helping to calculate the data transfer size and understand bandwidth usage in the forward direction. The total length of backward packets refers to the total length (in bytes) of all packets sent in the backward direction; this feature helps in calculating the data transfer size and understanding bandwidth usage in the backward direction. The maximum length of forwarding packets points out the maximum length of a single packet sent in the forward direction during the flow, helping to identify large packet transfers that can indicate certain types of applications. The minimum length of forwarding packets indicates the minimum length of a single packet sent in the forward direction during the flow, helping to identify small packet transfers that can aid in detecting certain types of control messages. Three transport protocol types, such as UDP, TCP, or Quick UDP Internet Connections (QUIC), are usually used by streaming applications. The target label represents different types of streaming traffic, 6 classes in this example, as follows. Apple apps widely use HTTP Live Streaming (HLS) for delivering media content. The Dynamic Adaptive Streaming over HTTP (DASH) protocol enables adaptive bitrate streaming with high-quality of media content from conventional HTTP web servers. Macromedia live streaming applications extensively use Real-Time Messaging Protocol (RTMP). WebRTC enables the sharing of audio, video, and data between browsers using JavaScript. RTP is used in conjunction with RTSP to facilitate the performing of audio and video streaming sessions, while MSS is widely employed in Microsoft's video streaming applications. (ii) Normalize the features using StandardScaler to improve the performance of the model. (iii) Train the DT, RF, and XGBoost models to learn to classify the application traffic. (iv) Evaluate the performance of the trained models with metrics such as precision, accuracy, and F1-score. (v) Once the models are trained and evaluated, they can be deployed to classify the streaming traffic in real-time. Listing 4.5 presents a Python code snippet for implementing the above-described traffic classification using DT, RF, and XGBoost. This code can be used as an application template for readers to address similar classification tasks.

```
#Import necessary libraries
import numpy as np
import pandas as pd
from sklearn.tree import DecisionTreeClassifier
from sklearn.ensemble import RandomForestClassifier
from xgboost import XGBClassifier
from sklearn.model_selection import train_test_split
from sklearn.preprocessing import StandardScaler
from sklearn.metrics import classification_report, accuracy_score, precision_score, f1_score

#Create a sample dataset
data = {
    'Flow.Duration': np.random.randint(100, 10000, 1000),
    'Total.Fwd.Packets': np.random.randint(1, 1000, 1000),
    'Total.Backward.Packets': np.random.randint(1, 1000, 1000),
    'Total.Length.of.Fwd.Packets': np.random.randint(100, 10000, 1000),
    'Total.Length.of.Bwd.Packets': np.random.randint(100, 10000, 1000),
    'Fwd.Packet.Length.Max': np.random.randint(50, 1500, 1000),
    'Fwd.Packet.Length.Min': np.random.randint(20, 1500, 1000),
    'Transport.Protocol': np.random.randint(0, 3, 1000),  # Assuming UDP, TCP, QUIC
    'label': np.random.randint(0, 6, 1000)  # Assuming HLS, DASH, RMTP, WebRTC, RTP, Microsoft_Smooth_Streaming
}

network_data = pd.DataFrame(data)

#Define the features and target
features = network_data[['Flow.Duration', 'Total.Fwd.Packets', 'Total.Backward.Packets',
                         'Total.Length.of.Fwd.Packets', 'Total.Length.of.Bwd.Packets',
                         'Fwd.Packet.Length.Max', 'Fwd.Packet.Length.Min', 'Transport.Protocol']]
target = network_data['label']

#Normalize the features
scaler = StandardScaler()
```

```
features_scaled = scaler.fit_transform(features)

#Split the data into training and testing sets
X_train, X_test, y_train, y_test = train_test_split(features_scaled,
    target, test_size=0.3, random_state=42)

#Initialize models
dt_model = DecisionTreeClassifier(random_state=42)
rf_model = RandomForestClassifier(n_estimators=100, random_state=42)
xgb_model = XGBClassifier(eval_metric='logloss')

#Train the models
dt_model.fit(X_train, y_train)
rf_model.fit(X_train, y_train)
xgb_model.fit(X_train, y_train)

#Evaluate the models
for model, name in zip([dt_model, rf_model, xgb_model], ['Decision Tree',
    'Random Forest', 'XGBoost']):
    y_pred = model.predict(X_test)

    # Compute accuracy, precision, and F1-score
    accuracy = accuracy_score(y_test, y_pred)
    precision = precision_score(y_test, y_pred, average='weighted')
    f1 = f1_score(y_test, y_pred, average='weighted')

    # Print evaluation metrics
    print(f"{name} Model:")
    print(f"Accuracy: {accuracy:.2f}")
    print(f"Precision: {precision:.2f}")
    print(f"F1-score: {f1:.2f}")
    print(classification_report(y_test, y_pred))

#Deploy the models for real-time classification of streaming traffic
#This step would involve integrating the trained models into the network
    monitoring system
```

Listing 4.5 Classification of Application Traffic

Evaluation metrics for classification models

The F1-score is another metric that can be used in conjunction with accuracy and precision (earlier discussed) to evaluate classification models. Overall, the F1-score assesses the balance between precision and recall (see Eq. 4.18); it is helpful in scenarios where identifying both false positives and false negatives has significant consequences. The Recall metric (see Eq. 4.17) measures the model's ability to identify TPs. Remarkably, a high value of recall means a few TPs were not detected. In turn, a low recall value suggests the model cannot detect a lot of TPs, resulting in more FNs. The Precision metric (see Eq. 4.13), as defined in this example and throughout the book, refers to the model's ability to predict anomalies accurately.

$$\text{Recall} = \frac{\text{TP}}{\text{TP} + \text{FN}} \tag{4.17}$$

$$F1 = 2 \cdot \frac{\text{Precision} \cdot \text{Recall}}{\text{Precision} + \text{Recall}} \tag{4.18}$$

4.3.5 Tree-Based Throughput Prediction

DT, RF, and XGBoost for network throughput prediction

ML algorithms such as DT, RF, and XGBoost can also be used to predict network throughput based on various factors (e.g., number of active users, type of application–video streaming and web browsing, and time of day)–ensuring efficient resource allocation and performance optimization. Following the methodology discussed in Chap. 3 and beyond the collection, data pre-processing, and feature engineering, it is necessary to (i) split the dataset into training and testing sets. (v) Train the DT, RF, and XGBoost models to predict throughput. (vi) Evaluate the performance of the trained models with MSE and R^2. (vii) Once the models are trained and evaluated, they can be deployed to predict the network traffic. Listing 4.6 shows a Python code snippet to realize the above-described throughput classification with DT, RF, and XGBoost. This code can serve as a template for readers to address similar classification tasks.

```
#Import necessary libraries
import numpy as np
import pandas as pd
from sklearn.model_selection import train_test_split
from sklearn.preprocessing import StandardScaler
from sklearn.metrics import mean_squared_error, r2_score
from sklearn.tree import DecisionTreeRegressor
from sklearn.ensemble import RandomForestRegressor
from xgboost import XGBRegressor

#Load and preprocess the data. Assume a dataset 'network_data' with
    features and target
data = {
    'active_users': [10, 20, 30, 40, 50, 60, 70, 80, 90, 100],
    'application_type': [1, 0, 1, 0, 1, 0, 1, 0, 1, 0],  # 1 for video
    streaming, 0 for other types
    'time_of_day': [9, 12, 15, 18, 21, 0, 3, 6, 9, 12],  # time in hours
    'network_throughput': [100, 150, 200, 250, 300, 350, 400, 450, 500,
    550]
}
network_data = pd.DataFrame(data)

#Define the independent features (X) and target (y)
X = network_data[['active_users', 'application_type', 'time_of_day']]
y = network_data['network_throughput']

#Normalize the features
scaler = StandardScaler()
features_scaled = scaler.fit_transform(X)

#Split the data into training and testing sets
X_train, X_test, y_train, y_test = train_test_split(features_scaled, y,
    test_size=0.3, random_state=42)

#Decision Tree Model
dt_model = DecisionTreeRegressor(random_state=42)
dt_model.fit(X_train, y_train)
y_pred_dt = dt_model.predict(X_test)
mse_dt = mean_squared_error(y_test, y_pred_dt)
r2_dt = r2_score(y_test, y_pred_dt)
print(f"Decision Tree - Mean Squared Error: {mse_dt:.2f}")
print(f"Decision Tree - R-squared: {r2_dt:.2f}")

#Random Forest Model
rf_model = RandomForestRegressor(n_estimators=100, random_state=42)
rf_model.fit(X_train, y_train)
y_pred_rf = rf_model.predict(X_test)
mse_rf = mean_squared_error(y_test, y_pred_rf)
r2_rf = r2_score(y_test, y_pred_rf)
print(f"Random Forest - Mean Squared Error: {mse_rf:.2f}")
print(f"Random Forest - R-squared: {r2_rf:.2f}")

#XGBoost Model
xgb_model = XGBRegressor(n_estimators=100, random_state=42)
xgb_model.fit(X_train, y_train)
y_pred_xgb = xgb_model.predict(X_test)
mse_xgb = mean_squared_error(y_test, y_pred_xgb)
r2_xgb = r2_score(y_test, y_pred_xgb)
print(f"XGBoost - Mean Squared Error: {mse_xgb:.2f}")
print(f"XGBoost - R-squared: {r2_xgb:.2f}")
```

Listing 4.6 Tree Regressors

4.4 Deep Neural Networks

An ANN that incorporates multiple neuron layers (i.e., hidden layers) between the input and output layers is typically referred to as a DNN. DNNs have become fundamental in various application domains, including image recognition, NLP, speech recognition, and network management. The input layer uses each neuron to receive the initial data; therefore, each neuron in this layer represents a feature of the dataset. For instance, in a resource allocation task, each neuron might represent the amount of memory a VNF is using. The hidden layers of neurons allow the modeling of complex relationships in the data; each neuron in a hidden layer receives input from the previous layer, computes weights that determine the strength of the connection between neurons, adds a bias that allows the model to fit the data more flexibly, and uses an activation function to produce an output that passes to the following layer. The output layer is responsible for carrying out the final prediction. Notably, the number of neurons in the output layer depends on the task. For a classification task, the output might be the type of server to use to allocate a VNF in the Cloud-Edge continuum. Remarkably, weights and biases are learned by the DNN autonomously during the training process. When training the DNN, network practitioners must set up hyperparameters, such as the number of hidden layers and the number of neurons per hidden layer.

Similar to supervised learning techniques previously discussed, with DNNs, network practitioners can, for instance, classify traffic in real-time to detect and mitigate security threats, or categorize it into types such as streaming, browsing, gaming, and downloading, to customize service delivery. Other network problems addressable by DNNs are (i) segment users based on their usage patterns and behaviors to define pricing schemes, (ii) classify network performance metrics to ensure the QoS is met, (iii) predict load patterns to balance network traffic dynamically, (iv) forecast demands to plan network growth; and (v) predict latency, throughput, and packet loss to meet SLAs.

4.4.1 Feedforward Neural Networks

An FNN (see Fig. 4.4) is a versatile type of DNN helpful in carrying out classification and regression tasks. FNNs comprise layers of neurons where each neuron in one layer connects to every neuron in the next layer. An FNN processes the input data through multiple layers and outputs the final prediction for management tasks such as identifying intrusions or predicting network throughput. FNNs are known for their adaptability, ability to learn complex patterns, and high performance.

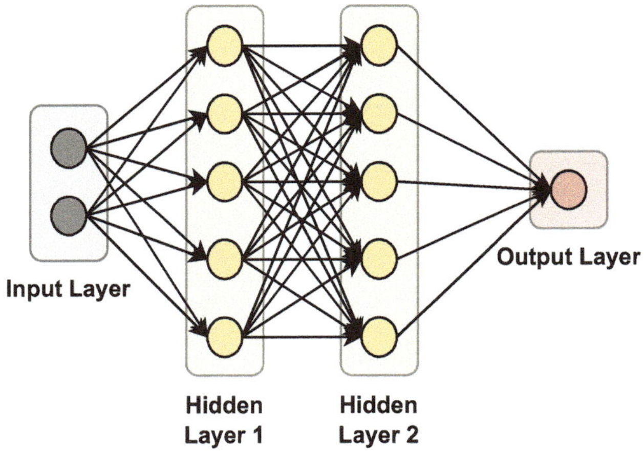

Fig. 4.4 Feedforward neural network

Algorithm 5 FNN Classifier

Input: Dataset D with features X and labels y, number of layers L
Output: FNN model F
Procedure:
Initialize the network with random weights and biases
for each epoch **do**
  for each batch of data **do**
    Forward propagate the input through the network to compute the output
    Compute the loss (error) between the predicted output and the actual label
    Backpropagate the error to update the weights and biases using gradient descent
  end for
end for
End Procedure

The FNN classifier (see Algorithm 5) includes:

- Initialize the FNN with random weights and biases.
- Iteratively improve the model by adjusting the weights and biases by forward propagation, computing the loss, and backpropagation.
  - Forward propagation. During forward propagation, the input data passes through the FNN layer by layer to produce the output. Each neuron behaves as follows. (i) It computes a weighted sum of its inputs. (ii) It adds a bias. (iii) It applies an activation function to introduce nonlinearity into the model, enabling the network to learn complex patterns. (iv) It passes the result of the activation function to the next layer; Sigmoid, ReLu, and Tanh are popular activation functions.
    - The Sigmoid activation function is an S-shaped curve that maps any real-valued number (x) to a value between 0 and 1; see Eq. 4.19. The gradient approaches zero for extremely large positive or negative inputs, leading to a slowdown in learning. i.e., the vanishing gradient problem. This activation function is typically used in the output layer of binary classification models to produce the probability of occurrence for each class.

$$\sigma(x) = \frac{1}{1 + e^{-x}}. \tag{4.19}$$

    - The ReLU activation function is computed by Eq. 4.20, where x is the value mapped to $[0, \infty)$. This activation function can lead to sparsity, as it outputs zero for negative inputs, which can improve computational efficiency. Remarkably, the gradient is 1 for a positive input and 0 for a negative input, avoiding the vanishing gradient problem. Furthermore, ReLU is widely used in hidden layers of DNNS due to its simplicity and effectiveness.

$$\text{ReLU}(x) = \max(0, x). \tag{4.20}$$

    - The Tanh activation function, given by Eq. 4.21, maps input values (x) to a range between -1 and 1. Unlike the Sigmoid function, Tanh activation is zero-centered, which can make optimization easier. The gradient approaches zero for an extremely large positive or negative input, which can also lead to the vanishing gradient problem. The Tanh activation function is often used in hidden layers of DNNs, mainly when the data is centered around zero.

$$\tanh(x) = \frac{e^x - e^{-x}}{e^x + e^{-x}}. \tag{4.21}$$

  - Computing loss. A loss function measures the error between the predicted output and the actual label, guiding the optimization process. Common loss functions include Mean Squared Error (MSE) for regression tasks and Cross-Entropy Loss for classification tasks. Cross-entropy extends the concept of entropy by measuring the difference between two probability distributions. For binary classification, the cross-entropy loss is given by Eq. 4.22. Where N is the number of samples, y_i is the actual label, and $\hat{y}_i$ is the predicted probability of the positive class. Equation 4.23 shows the categorical cross-entropy loss for C classes. Where $y_{i,c}$ is a binary indicator (0 or 1) if class label c is the correct classification for sample i, and $\hat{y}_{i,c}$ is the predicted probability of sample i being in class c.

$$\text{Binary Cross-Entropy} = -\frac{1}{N}\sum_{i=1}^{N}\left[y_i \log(\hat{y}_i) + (1 - y_i)\log(1 - \hat{y}_i)\right] \tag{4.22}$$

$$\text{Categorical Cross-Entropy} = -\frac{1}{N}\sum_{i=1}^{N}\sum_{c=1}^{C} y_{i,c}\log(\hat{y}_{i,c}). \tag{4.23}$$

– Backpropagation. During backward propagation, the network computes the gradient of the loss function with respect to each weight and bias. This computation is performed using the chain rule. Once the gradients are calculated, an optimization algorithm adjusts the weights and biases to minimize the loss function. SGD, RMSprop, and Adam are commonly used optimization algorithms.

  · SGD is characterized by using a single training example to compute the gradient and update the parameters of DNNs. This optimization algorithm is fast and can handle large datasets and online scenarios, but it introduces high variance in updates. SGD operates as follows. (i) Initialize Parameters. (ii) Randomly shuffle the training data. (iii) Iterate over training samples: compute the gradient of the loss function for the current training sample and update the parameters. (iii) Repeat steps (ii) and (iii) for multiple epochs until the loss function converges. The gradient of the loss function (J, see Eq. 4.24), evaluated at the $i$ith training sample, measures the difference between the predicted output of the model $y^{(i)}$ and the actual target value $x^{(i)}$ concerning the parameters (θ: weights and biases) of the model. The updating of parameters given by Eq. 4.25 indicates that the new parameters θ are obtained by subtracting the product of the learning rate η and the gradient $\nabla J(\theta; x^{(i)}, y^{(i)})$ from the current parameters θ. This update shifts the parameters in a direction that reduces the loss function, thereby improving the model.

$$\nabla J(\theta; x^{(i)}, y^{(i)}) = \left.\frac{\partial J(\theta)}{\partial \theta}\right|_{x^{(i)}, y^{(i)}} \tag{4.24}$$

$$\theta = \theta - \eta \nabla J(\theta; x^{(i)}, y^{(i)}). \tag{4.25}$$

  · RMSprop is characterized by an adaptive learning rate and the prevention of exploding gradients (i.e., the gradients used to update the DNN's weights grow exponentially). This optimization algorithm operates as follows. (i) Initialize Parameters. (ii) Compute the gradient of the loss function for the current training sample. (iii) Update moving average by using Eq. 4.26, where γ is the decaying rate and $E[g^2]_t$ is the exponentially weighted average of the squared gradients. (iv) Scale the gradient, see Eq. 4.27 where η is the learning rate, θ_t are the parameters at time step t, and ϵ is a small constant to prevent division by zero. (v) Repeat steps (ii) to (iv) for multiple epochs until the loss function converges.

$$E[g^2]_t = \gamma E[g^2]_{t-1} + (1 - \gamma)(\nabla J(\theta; x^{(i)}, y^{(i)}))^2 \tag{4.26}$$

$$\theta_t = \theta_{t-1} - \frac{\eta}{\sqrt{E[g^2]_t + \epsilon}} \nabla J(\theta; x^{(i)}, y^{(i)}). \tag{4.27}$$

  · Adam combines RMSprop and Momentum (i.e., an SGD extension designed to overcome the oscillations of noisy gradients and avoid getting stuck in flat spots of the search space) to provide an adaptive learning rate for each parameter, making it efficient and robust for training deep neural networks. This optimization algorithm, widely used in practice for many models, operates as follows. (i) Initialize Parameters. (ii) Calculate the gradient of the loss function. (iii) Calculate first (m_t) and second v_t moment, see Eqs. 4.28 and 4.29, where β_1 and β_2 are the exponential decay rate for the first and second moment that controls how much of the past gradients to consider while updating m_t and v_t, m_{t-1} is the first moment of the previous time step. (iv) Correct the first and second moment; see Eqs. 4.30 and 4.31. (v) Update parameters according to Eq. 4.32, where ϵ represents a small constant added to prevent division by zero in the update rule. (vi) Repeat steps (ii) to (v) until convergence. Remarkably, Adam is computationally expensive and sensitive to hyperparameters.

$$m_t = \beta_1 m_{t-1} + (1 - \beta_1)\nabla J(\theta; x^{(i)}, y^{(i)}) \tag{4.28}$$

$$v_t = \beta_2 v_{t-1} + (1 - \beta_2)(\nabla J(\theta; x^{(i)}, y^{(i)}))^2 \tag{4.29}$$

$$\hat{m}_t = \frac{m_t}{1 - \beta_1^t} \tag{4.30}$$

$$\hat{v}_t = \frac{v_t}{1 - \beta_2^t} \tag{4.31}$$

$$\theta_t = \theta_{t-1} - \frac{\eta}{\sqrt{\hat{v}_t} + \epsilon}\hat{m}_t. \tag{4.32}$$

- The final model consists of optimized weights and biases that minimize the loss function, providing accurate predictions for new input data.
- FNNs are versatile and can model complex, nonlinear relationships in the data by adjusting the architecture. i.e., the number of layers, neurons, and hyperparameters. Despite their power, FNNs are challenging to interpret and require significant computational resources for training, especially with large datasets and deep architectures. They are also sensitive to hyperparameter tuning, requiring careful selection of learning rates, batch sizes, and network architectures for optimal performance. Training FNNs can lead to issues like vanishing or exploding gradients, which can hinder convergence; techniques such as normalization, dropout, and advanced initialization methods are often used to address these challenges.

4.4.2 Convolutional Neural Networks

A CNN (see Fig. 4.5) is a powerful type of DNN comprising multiple layers, designed to automatically and adaptively learn spatial hierarchies of features from (mainly) images or other types of input data. A CNN processes the input data through convolutional, pooling, and fully connected layers to output the final prediction for classification or regression tasks. CNNs are well-known for their high performance and capability to learn complex patterns directly from raw data. In this sense, in network management, they have been widely used to predict available bandwidth, estimate QoS parameters, and classify encrypted traffic.

The CNN classifier (see Algorithm 6) includes:

- Initialize the CNN with random weights and biases.
- Iteratively improve the model by adjusting the weights and biases through forward propagation, computing the loss, and backpropagation.
  - During forward propagation, the input data passes through the network layer by layer to produce the output. Each layer in a CNN typically includes convolutional layers, activation functions, and pooling layers.

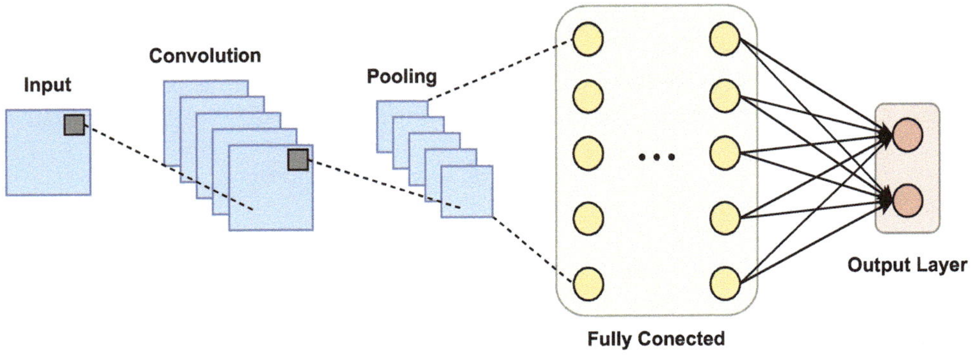

Fig. 4.5 Convolutional neural network

Algorithm 6 CNN Classifier

Input: Dataset D with features X and labels y, number of layers L
Output: CNN model C
Procedure:
Initialize the network with random weights and biases
for each epoch **do**
  for each batch of data **do**
    Forward propagate the input through the convolutional layers,
    activation functions, pooling layers, and fully connected layers to compute the output
    Compute the loss (error) between the predicted output and the actual label
    Backpropagate the error to update the weights and biases using gradient descent
  end for
end for
End Procedure

- A convolutional layer is designed to automatically and adaptively learn spatial hierarchies of feature maps by convolving the input data with filters (kernels). A convolutional operation works as follows. The filter g is placed on the input data f such that the top-left corner of the filter aligns with the position (x, y) in the input data. For each position of the filter, the values of the input data f that overlap with it are multiplied element-wise with the corresponding values of g. The resulting products are summed to produce a single value, which is then assigned to the corresponding position (x, y) on the output feature map. The filter is then moved (slid) to the next position in the input data, and the process is repeated until all of the input data has been covered. Equation 4.33 offers a 2D convolution operation. Where $(f * g)(x, y)$ represents the result of the convolutional operation at position (x, y) in the output feature map. $\sum_{i=0}^{m-1} \sum_{j=0}^{n-1}$ are the summation operations on the indices i and j, which range from 0 to $m - 1$ and 0 to $n - 1$ respectively. Here, m and n are the height and width of the filter. $g(i, j)$ is the value of the filter g at position (i, j). $f(x - i, y - j)$ is the value of the input data f at position $(x - i, y - j)$. The function f represents the input feature map from the previous layer.

$$(f * g)(x, y) = \sum_{i=0}^{m-1} \sum_{j=0}^{n-1} g(i, j) \cdot f(x - i, y - j) \tag{4.33}$$

- As in FNNs, activation functions in CNNs introduce nonlinearity into the model, enabling the network to learn complex patterns; typical activation functions include ReLU, Sigmoid, and Tanh.

- A pooling layer reduces the spatial dimensions of the feature, lowering parameters and computation in the CNN. A typical pooling method is given by Equation 4.34, where $\max_{i=0}^{p-1} \max_{j=0}^{q-1}$ denotes the nested maximization operations on the indices i and j, which range from 0 to $p - 1$ and 0 to $q - 1$ respectively. In turn, p and q are the height and width of the pooling window. $h(x + i, y + j)$ is the value of the input feature map h at position $(x + i, y + j)$. The function h represents the activations from the previous layer.

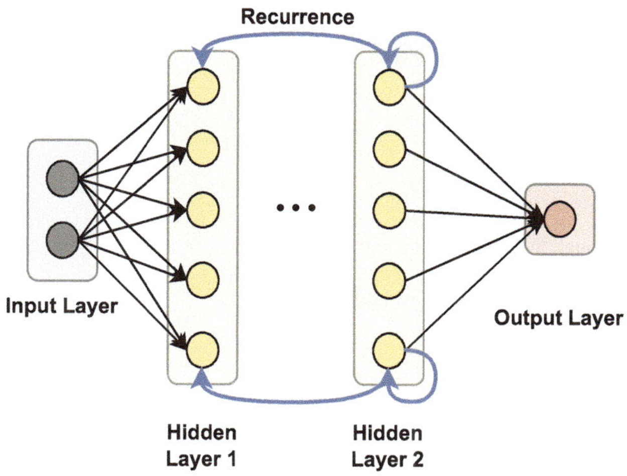

Fig. 4.6 Recurrent neural network

$$\text{MaxPool}(x, y) = \max_{i=0}^{p-1} \max_{j=0}^{q-1} h(x + i, y + j) \tag{4.34}$$

- Calculating loss using, for example, the Eqs. 4.22 and 4.23 for classification tasks.
- As in FNNs, during backward propagation, a CNN computes the gradient of the loss function with respect to each weight and bias. Once the gradients are calculated, an optimization algorithm, such as Adam, adjusts the weights and biases to minimize the loss function.

- The final model consists of optimized weights and biases that minimize the loss function, providing accurate predictions for new input data.
- CNNs require significant computational resources for training. Furthermore, they are sensitive to hyperparameter tuning. Remarkably, data augmentation, dropout, and batch normalization are techniques often employed to enhance the performance and generalization of CNN models.

4.4.3 Recurrent Neural Networks

An RNN (see Fig. 4.6) is a DNN designed to perform classification and regression tasks on sequential data. Unlike FNNs, RNNs have connections that form directed cycles, allowing information to persist across different sequence steps. This capability makes RNNs particularly effective for tasks where context and order are crucial, such as NLP, time series forecasting, and speech recognition. In network management, they have primarily helped in predicting future bandwidth utilization to optimize network capacity planning, thereby avoiding congestion. Additionally, they have enabled the classification of different types of network traffic by learning from historical data, which facilitates better management of resource allocation and QoS.

The RNN (see Algorithm 7) includes:

- Initialize the RNN with random weights and biases.
- Iteratively improve the model by adjusting the weights and biases through forward propagation, computing the loss, and BackPropagation Through Time (BPTT).

Algorithm 7 RNN Classifier

Input: Sequential dataset D with features X and labels y, number of time steps T
Output: RNN model R
Procedure:
Initialize the network with random weights and biases
for each epoch **do**
  for each sequence in the data **do**
    for each time step t from 1 to T **do**
      Forward propagate the input through the network to compute the output
      Compute the loss (error) between the predicted output and the actual label
      Backpropagate the error through time to update the weights and biases
    end for
  end for
end for
End Procedure

  - During forward propagation, the input data passes through the network one time step at a time. The RNN maintains a hidden state at each step, capturing information from previous steps. This information is combined with the current input to generate the output and update the hidden state. Equation 4.35 shows the forward propagation step, where h_t and x_t are the hidden state and input at time step t, W and U are weight matrices, and b represents the bias vector. Furthermore, per time step, each neuron applies an activation function, such as ReLU, and passes the result to the next layer and the next time step.

$$h_t = \sigma(W \cdot h_{t-1} + U \cdot x_t + b). \tag{4.35}$$

  - Computing Loss. Typical loss functions for sequential tasks include MSE for regression and Cross-Entropy for classification.
  - During BPTT, the RNN computes the gradient of the loss function with respect to each weight and bias through the sequence of time steps using the chain rule. Once these gradients are calculated, an optimization algorithm, such as SGD, RMSprop, or Adam, adjusts the weights and biases to minimize the loss function.

- The final model consists of optimized weights and biases that minimize the loss function, providing accurate predictions for new sequential data.
- Gradient clipping, advanced initialization methods, and more sophisticated RNNs, such as LSTM and GRU, are techniques that help in addressing the vanishing and exploding gradient issues, which can hinder convergence and make the training of RNNs challenging. Another challenge when using RNNs is that they require significant computational resources for training, especially for long sequences and deep architectures.

4.4.4 Encrypted Traffic Classification

CNN for encrypted traffic classification
For network performance or security management, a CNN model can be used to classify encrypted traffic. In this regard, such a model enables the detection of threats, optimizes network resources, enforces policies, and provides visibility into application usage without requiring the decryption of content, thereby preserving both privacy and security. Following the methodology discussed in Chap. 3 and beyond the collection, data pre-processing, and feature engineering, it is necessary to (i) create a dataset of example with features including $packet\_size$, $flow\_duration$, $bytes\_sent$, $bytes\_received$, and $protocol\_type$ e.g., Transport Layer Security (TLS) protocol is a widely adopted to provide privacy and data security, MTProto is the messaging protocol used by Telegram to perform end-to-end-encryption, and Signal Protocol provides to Signal Messaging App end-to-end encryption for voice and instant chats. The target label represents different types of encrypted traffic (five classes: WhatsApp, Telegram, iMessage, Facebook_Messenger, and Signal). (ii) Normalize the features

using StandardScaler to improve the CNN's performance. (iii) Reshape the data to fit the input the CNN requires (samples, timesteps, features). (iv) Build the CNN model, including a Conv1D layer with 64 filters, a MaxPooling1D layer, a Flatten layer, and two Dense layers. The first Dense layer has 64 neurons with ReLU activation, and the output layer has neurons equal to the number of classes with Softmax activation. (v) Compile the model with the Adam optimizer and categorical cross-entropy loss function. (vi) Training the model for 50 epochs with a batch size of 32 and a validation split of 20%. (vii) Evaluate the model's performance using a classification report, including metrics like precision, recall, and F1-score for each class. (vii) Once the models are trained and evaluated, they can be deployed to classify the encrypted network traffic. Listing 4.7 shows a Python code snippet to realize the above-described encrypted traffic classification with a CNN. Readers can use this code as a template to address similar classification problems.

```
#Import necessary libraries
import numpy as np
import pandas as pd
from sklearn.model_selection import train_test_split
from sklearn.preprocessing import StandardScaler
from sklearn.metrics import classification_report
from tensorflow.keras.models import Sequential
from tensorflow.keras.layers import Conv1D, MaxPooling1D, Flatten, Dense
from tensorflow.keras.utils import to_categorical

#Load and preprocess the data
#Assume a dataset 'network_data' with features and target
#For demonstration purposes, create a sample dataset
data = {
    'packet_size': np.random.rand(1000),
    'flow_duration': np.random.rand(1000),
    'protocol_type': np.random.randint(0, 3, 1000),  # Assuming 3
    different protocols
    'bytes_sent': np.random.rand(1000),
    'bytes_received': np.random.rand(1000),
    'label': np.random.randint(0, 5, 1000)  # Assuming 5 different types
    of encrypted traffic
}
network_data = pd.DataFrame(data)

#Define the features and target
features = network_data[['packet_size', 'flow_duration', 'protocol_type',
    'bytes_sent', 'bytes_received']]
target = network_data['label']

#Normalize the features
scaler = StandardScaler()
features_scaled = scaler.fit_transform(features)

#Reshape the data for CNN (3D shape: samples, timesteps, features)
X = np.expand_dims(features_scaled, axis=2)

#Convert labels to categorical (one-hot encoding)
y = to_categorical(target)

#Split the data into training and testing sets
X_train, X_test, y_train, y_test = train_test_split(X, y, test_size=0.3,
    random_state=42)

#Build the CNN model
model = Sequential()
model.add(Conv1D(64, kernel_size=2, activation='relu', input_shape=(
    X_train.shape[1], 1)))
model.add(MaxPooling1D(pool_size=2))
model.add(Flatten())
model.add(Dense(64, activation='relu'))
model.add(Dense(y_train.shape[1], activation='softmax'))

#Compile the model
```

```
model.compile(optimizer='adam', loss='categorical_crossentropy', metrics
    =['accuracy'])

#Train the model
model.fit(X_train, y_train, epochs=50, batch_size=32, validation_split
    =0.2)

#Evaluate the model
y_pred = model.predict(X_test)
y_pred_classes = np.argmax(y_pred, axis=1)
y_test_classes = np.argmax(y_test, axis=1)

#Print report that includes metrics like precision, recall, and F1-score
    for each class.
print(classification_report(y_test_classes, y_pred_classes))
```

Listing 4.7 CCN multi-classifier

4.4.5 Estimating Bandwidth Utilization in 6G

RNN for predicting bandwidth utilization in 6G networks

For network performance and accounting management, an RNN can help predict bandwidth utilization to optimize network capacity planning and avoid congestion by learning from historical data, enabling better network resource scheduling as well as meeting QoS and QoE requirements. Following the methodology discussed in Chap. 3 and beyond the collection, data pre-processing, and feature engineering, it is necessary to (i) create a sample dataset with historical bandwidth utilization that considers the type of user (i.e., gold, silver, and bronze) and 6G use case: ultrareliable low-latency communication (uRLLC—services needing ultrahigh reliability and extremely low latency), massive machine-type communication (mMTC—services requiring the connectivity of a wide gamma of devices), enhanced mobile broadband (eMBB—services demanding high data traffic and a bit rate of 20 Gbps and 100 Mbps), strengthened enhanced mobile broadband (sEMBB embraces EMBB services demanding high QoE), ultramassive machine-type communications (umMTC—services needing a much more massive number of simultaneous connections per space than mMTC), massive ultrareliable low-latency communications (mURLLC—mission-critical services requiring high reliability, low latency, and high availability), mobile broadband reliable low-latency communications (MBRLLC—services demandin uRLLC and eMBB), and extremely reliable and low-latency communications (ERLLC—services that merge URLLC and mMTC demands). (ii) Normalize the $bandwidth\_utilization$ using MinMaxScaler. (iii) Encoding categorical features ($user\_type$ and use_case) using LabelEncoder. (iv) Create sequence data by combining the normalized and encoded features. (v) Building the RNN model, including a SimpleRNN layer with 50 units and Dense layers for prediction. (vi) Compile the model with the Adam optimizer and MSE loss function. (vi) Training the model for 50 epochs with a batch size of 32 and a validation split of 20%. (vii) Evaluate the model's performance using MSE. The results are plotted to compare the actual and predicted bandwidth utilization values. Listing 4.8 shows a Python code snippet to realize the above-described bandwidth consumption prediction using an RNN. Readers can use this code as a template for addressing similar regression problems.

```
#Import necessary libraries
import numpy as np
import pandas as pd
from sklearn.model_selection import train_test_split
from sklearn.preprocessing import MinMaxScaler, LabelEncoder
from sklearn.metrics import mean_squared_error
import matplotlib.pyplot as plt
from tensorflow.keras.models import Sequential
from tensorflow.keras.layers import SimpleRNN, Dense, Input, concatenate
from tensorflow.keras.models import Model

#Load and preprocess the data
#Assume a dataset 'network_data' with historical bandwidth utilization
#For demonstration purposes, create a sample dataset
np.random.seed(42)
data = {
    'timestamp': pd.date_range(start='1/1/2020', periods=1000, freq='H'),
```

```
    'bandwidth_utilization': np.random.rand(1000) * 100,  # Random
    bandwidth utilization values
    'user_type': np.random.choice(['silver', 'gold', 'bronze'], size
    =1000),
    'use_case': np.random.choice(['embb', 'mmtc', 'urllc','sembb','ummtc'
    ,'murllc','mbrllc','erllc'], size=1000)}

network_data = pd.DataFrame(data)

#Sort data by timestamp
network_data = network_data.sort_values('timestamp')

#Normalize the bandwidth utilization
scaler = MinMaxScaler(feature_range=(0, 1))
network_data['bandwidth_utilization'] = scaler.fit_transform(network_data
    [['bandwidth_utilization']])

#Encode categorical features
label_encoder_user = LabelEncoder()
network_data['user_type_encoded'] = label_encoder_user.fit_transform(
    network_data['user_type'])

label_encoder_use_case = LabelEncoder()
network_data['use_case_encoded'] = label_encoder_use_case.fit_transform(
    network_data['use_case'])

#Prepare the data for RNN
def create_sequences(data, seq_length):
    xs, ys = [], []
    for i in range(len(data) - seq_length):
        x = data[i:i+seq_length]
        y = data[i+seq_length, 0]  # Predicting the '
    bandwidth_utilization'
        xs.append(x)
        ys.append(y)
    return np.array(xs), np.array(ys)

#Combine features into a single array
features = network_data[['bandwidth_utilization', 'user_type_encoded', '
    use_case_encoded']].values
seq_length = 10  # Sequence length
X, y = create_sequences(features, seq_length)

#Split the data into training and testing sets
X_train, X_test, y_train, y_test = train_test_split(X, y, test_size=0.2,
    random_state=42)

#Separate features for RNN input
X_bandwidth_train = X_train[:, :, 0].reshape((X_train.shape[0], X_train.
    shape[1], 1))
X_user_type_train = X_train[:, :, 1]
X_use_case_train = X_train[:, :, 2]

X_bandwidth_test = X_test[:, :, 0].reshape((X_test.shape[0], X_test.shape
    [1], 1))
X_user_type_test = X_test[:, :, 1]
X_use_case_test = X_test[:, :, 2]

#Build the RNN model
input_shape = (seq_length, 1)
bandwidth_input = Input(shape=input_shape, name='bandwidth_input')
rnn_out = SimpleRNN(50, activation='relu')(bandwidth_input)

user_input = Input(shape=(seq_length,), name='user_input')
use_case_input = Input(shape=(seq_length,), name='use_case_input')

```

```python
merged = concatenate([rnn_out, user_input, use_case_input])
dense_out = Dense(10, activation='relu')(merged)
output = Dense(1)(dense_out)

model = Model(inputs=[bandwidth_input, user_input, use_case_input],
    outputs=output)

#Compile the model
model.compile(optimizer='adam', loss='mean_squared_error')

#Train the model
history = model.fit([X_bandwidth_train, X_user_type_train,
    X_use_case_train], y_train, epochs=50, batch_size=32,
    validation_split=0.2, verbose=1)

#Evaluate the model
y_pred = model.predict([X_bandwidth_test, X_user_type_test,
    X_use_case_test])
y_pred_inverse = scaler.inverse_transform(y_pred)
y_test_inverse = scaler.inverse_transform(y_test.reshape(-1, 1))

#Calculate mean squared error
mse = mean_squared_error(y_test_inverse, y_pred_inverse)
print(f"Mean Squared Error: {mse:.2f}")

#Plot the results
plt.figure(figsize=(12, 6))
plt.plot(y_test_inverse, label='True Bandwidth Utilization')
plt.plot(y_pred_inverse, label='Predicted Bandwidth Utilization')
plt.title('True vs Predicted Bandwidth Utilization')
plt.xlabel('Time')
plt.ylabel('Bandwidth Utilization')
plt.legend()
plt.show()
```

Listing 4.8 RNN regressor

4.5 Summary

This chapter initially introduced various supervised ML algorithms, ranging from simple linear regression to RNN. Furthermore, it exemplified the use of several ML models for regression and classification tasks from a practical perspective (code snippets in Python were presented) and discussed how supervised ML supports various management tasks: network latency prediction, network throughput prediction, anomaly detection, encrypted traffic classification, and bandwidth utilization foreseeing.

Homework Problems and Questions

4.1 Define and exemplify classification tasks in network management. Also, raise candidate algorithms and evaluation metrics.

4.2 Define and exemplify a regression task in network management. Also, raise candidate algorithms and evaluation metrics.

4.3 Define and exemplify a regression task in network management. Also, raise candidate algorithms and evaluation metrics.

4.4 Use polynomial regression with Ridge regularization to predict network traffic (target y) by considering it may exhibit daily peaks and troughs (independent feature X).

4.5 What are the differences and similarities of Ridge and Lasso regularization?.

4.6 Use SVM, XGBoost, CNN, and RNN to classify streaming application traffic. The dataset includes as features to flow duration, the total of forwarding packets, the total backward of packets, the total length of forwarding packets, the total length of backward packets, the maximum length of forwarding packets, the minimum length of forwarding packets, and the transport protocol (UDP, TCP, or QUIC). The target label represents different types of streaming traffic, 6 classes: HLS, DASH, RMTP, WebRTC, RTP, and MSS. Compare the performance of the trained models regarding precision, accuracy, balanced accuracy, and F1-score.

4.7 How do MinMaxScaler, StandarScaler, and RobustScaler operate? When is it appropriate to use them?.

4.8 What are the differences between accuracy and balanced accuracy? When is it appropriate to use balance accuracy and F1-Score?.

Solutions of This Chapter

4.1 Problem. Define and exemplify classification tasks in network management. Also, raise candidate algorithms and evaluation metrics.

Solution. A binary or multi-classification task refers to foreseeing a discrete variable. A binary variable relates to detecting anomalous traffic while detecting the type of traffic refers to multi-classification. In this regard, detecting malware involves binary classification, whereas determining the type requires multi-classification. Candidate algorithms for classification tasks are Logistic Regression, Classification and Regression Trees, Random Forest, Gradient Boosting Machines (e.g., XGBoost, LightGBM, CatBoost), SVM, CNN, FNN, RNN, k-NN, Naive Bayes, Ensemble Methods, AEs, VAEs, and transformers. Metrics for evaluating those algorithms are accuracy, precision, recall, f1-score, Compute Area Under the Receiver Operating Characteristic Curve, Confusion Matrix, and MCC.

4.2 Problem. Define and exemplify regression tasks in network management. Also, raise candidate algorithms and evaluation metrics.

Solution. A regression task refers to foresee a continuous variable. Essential metrics to be predicted with regression algorithms are delay, packet loss, throughput, and reliability; remarkably, these metrics are critical to handling QoS autonomously. Delay is the total time a packet spends going from the source to the destination, involving propagation, transmission, processing, and queuing delays. Packet loss is the ratio of packets that fail to reach their destination due to, for instance, network congestion, faulty hardware, or transmission errors. Throughput is the amount of data successfully transmitted from source to destination per unit time (e.g., Mbps). Reliability is the ability of the network to deliver data without errors, loss, or interruptions. Candidate algorithms for regression tasks are Ridge Regression, Lasso Regression, Classification and Regression Trees, Random Forest, Gradient Boosting Machines (e.g., XGBoost, LightGBM, and CatBoost), Support Vector Regressor, CNN, RNN, FNN, Gaussian Process Regression, Bayesian Regression, KNN Regressor, Ensemble Methods (e.g., Stacking, Bagging, and Boosting). Metrics for evaluating those algorithms are MAE, MSE, RMSE, R-squared, Adjusted R-squared, MAPE, and Explained Variance.

4.3 Problem. Use polynomial regression with Ridge regularization to predict network traffic (target y) by considering it may exhibit daily peaks and troughs (independent feature X).

Solving process. Follow the methodology defined in Chap. 3.

- Business and data understanding. This problem aims to predict network traffic quantity. The data is generated using the Numpy library, creating a data frame with the feature (X, time of day) (here, we can use the Numpy random method to generate the values) and its target (y, network traffic).

- Data engineering.
  - Import LinearRegression and PolynomialFeatures libraries.
  - Remove the duplicates and empty instances.
  - Convert the X feature to a Polynomial Feature.
  - Normalizing (it is crucial to improving model prediction) with the MinMaxScaler converts polynomial feature values into values between 0 and 1.
  - Split the normalized data into train and test sets; a typical split is 80:20, 80% for training, and 20% for testing.
- ML model engineering.
  - Create a Ridge-based regression model and train it using the polynomial features.
  - Use the trained model to make predictions on the testing set.
- Quality assurance for ML applications.
  - Evaluate the model's performance using RMSE and R^2.
  - Print the evaluation metrics and visualize the actual data and the predicted trend.
  - Discuss the results.

4.4 Problem. What are the differences and similarities of Ridge and Lasso regularization?.

Solution. Ridge Regularization adds as a penalty to the loss function the squared magnitude of the coefficients, considers all features in the model (even if their contribution is small), and consequently, operates well when most features are relevant. Lasso adds as a penalty the absolute magnitude of the coefficients, shrinks some coefficients to zero, and, as a result, performs feature selection that is useful for models with many features where only a subset is relevant. Remarkably, Ridge presents a lower computational complexity than Lasso for large datasets, and the latter is more robust to outliers than the former.

4.5 Problem. Use SVM, XGBoost, CNN, and RNN to classify streaming application traffic. The dataset includes as features to flow duration, the total of forwarding packets, the total backward of packets, the total length of forwarding packets, the total length of backward packets, the maximum length of forwarding packets indicates the maximum length of a single packet sent in the forward direction during the flow, the minimum length of forwarding packets, and the transport protocol (UDP, TCP, or QUIC). The target label represents different types of streaming traffic, 6 classes: HLS, DASH, RMTP, WebRTC, RTP, and MSS. Compare the performance of the trained models regarding precision, accuracy, balanced accuracy, and F1-score.

Solving process. Following the methodology discussed in Chap. 3.

- Business and data understanding. The goal is to classify streaming application traffic.
  - Create a dataset with target (streaming traffic: HLS, DASH, RMTP, WebRTC, RTP, and MSS) and features (flow duration, the total of forwarding packets, the total backward of packets, the total length of forwarding packets, the total length of backward packets, the maximum length of forwarding packets, the minimum length of forwarding packets, and the transport protocol (UDP=17, TCP=6, and QUIC=443).
  - Generate the data using the Numpy library, creating a data frame using the Numpy random method to generate the values for the features.
- Data engineering.
  - Remove the duplicates and empty instances and searches for atypical values to remove them.
  - Normalize the data frame using RobustScaler, which is appropriate for datasets with outliers.
  - Split into train and test sets the normalized data; a typical split is 80:20, 80% for training and 20% for testing.

- ML model engineering.

  - Build the SVM-based model with the RBF kernel.
  - Create the XGBoost-based model, perform testing to get the optimal learning rate, max depth, and number of estimators.
  - Build the RNN-based model using LSTM, with the last layer activated by Softmax (commonly used for multi-class classification) and the loss function CategoricalCrossentropy.
  - Create the CNN-based model using a Conv1D layer, with the Softmax activation function in the last layer and the loss function CategoricalCrossentropy.

- Quality assurance for ML applications.

  - Evaluate each model, based on SVM, XGBoost, RNN, and CNN, using the test set regarding accuracy, balanced accuracy, precision, and F1-score.
  - Plotted the evaluation results using the Matplotlib library.
  - Discuss the results.

4.6 Problem. How do MinMaxScaler, StandarScaler, and RobustScaler operate? When is it appropriate to use them?.

Solution. Feature scaling is a crucial step in data engineering. In this regard, it is essential to visualize your data with histograms or boxplots before choosing a scaler. Furthermore, it is suggested that different scalers be experimented with to evaluate model performance and make an optimal choice. Consider scaling different features with diverse methods for datasets with both Gaussian and non-Gaussian features.

- MinMaxScaler scales features to a fixed range, typically [0, 1] according to Eq. 4.36. Where X is the original feature value and X_{min} and X_{max} are their minimum and maximum values. This scaling method is sensitive to outliers and suitable when the dataset has a bounded range (e.g., pixel intensities in images) and data non follows a Gaussian distribution and for algorithms like KNN that rely on distance metrics.

$$X_{\text{scaled}} = \frac{X - X_{\text{min}}}{X_{\text{max}} - X_{\text{min}}}. \tag{4.36}$$

- StandardScaler scales features to obtain data with a mean of 0 and a standard deviation of 1, see Eq. 4.37. Where μ and σ refer to the feature's mean and standard deviation. This scaling method does not guarantee a specific range for the scaled data and is appropriate when the data follows a Gaussian distribution and algorithms (linear regressors and PCA) assume normally distributed data. Remarkably, StandardScaler is less sensitive to outliers than MinMaxScaler.

$$X_{\text{scaled}} = \frac{X - \mu}{\sigma}. \tag{4.37}$$

- RobustScaler scales features using the median and interquartile range to reduce the influence of outliers; see Eq. 4.38. Where Median is the median of the feature and IQR is the interquartile range ($Q3 - Q1$). This scaling method does not guarantee a specific range for the scaled data and is suitable when the data contains many outliers and does not follow a Gaussian distribution.

$$X_{\text{scaled}} = \frac{X - \text{Median}}{\text{IQR}}. \tag{4.38}$$

4.7 Problem. What are the differences between accuracy and balanced accuracy? When is it appropriate to use balance accuracy and F1-Score?.

Solution—Question 1. Accuracy measures the proportion of correctly predicted instances of the total samples. This metric is suitable when the dataset is balanced: in this scenario, it allows giving equal importance to positive and negative predictions.

Balanced accuracy is appropriate when dealing with imbalanced datasets since it averages the recall (sensitivity) of each class. This metric ensures that positive and negative predictions are given equal importance, regardless of the classes size in the dataset.

Solution—Question 2. As mentioned, balanced accuracy is appropriate when the dataset is imbalanced. The aim is to evaluate the model's performance in all classes with the same importance. F1-Score is also suitable for imbalanced datasets, and the goal is to focus on assessing the performance of one class.

Network Management Driven by Unsupervised Learning

5

5.1 Clustering

With clustering algorithms, network practitioners can, for instance, (i) cluster network traffic data into normal and abnormal patterns to identify intrusions; (ii) group traffic data into clusters, such as streaming, browsing, gaming, and downloading, to understand the composition of network traffic, aiming at optimizing resource allocation based on traffic types; (iii) segment network users based on their usage patterns and behaviors to perform personalized service offerings, targeted marketing, and network capacity planning; (iv) group historical fault data into different categories of faults to help diagnose and predict network issues, enable proactive maintenance, and reduce downtime; (v) cluster QoS metrics, such as latency, packet loss, and throughput, to ensure that all groups of users receive the required service quality and identify network areas needing improvement; (vi) group traffic data to identify load patterns that can be used to balance the load dynamically, preventing any single server or link from becoming a bottleneck; and (vii) cluster historical usage data to identify trends and patterns to forecast future demands and plan network expansions and upgrades accordingly.

5.1.1 K-Means

K-means (see Fig. 5.1) is an extremely popular unsupervised ML algorithm that is well-known for its efficiency to carry out clustering tasks.hyper@itemfalse This clustering algorithm splits the dataset into k distinct, non-overlapping groups or classes. K-means is widely used in network management of performance or security for anomaly detection, traffic analysis, and user behavior segmentation tasks.

Algorithm 1 K-means clustering

Input: Dataset D with data points X, number of clusters K
Output: Cluster centroids $\mu_1, \mu_2, \ldots, \mu_K$
Procedure:
Initialize K centroids randomly
repeat
  Assign each data point to the nearest centroid to form K clusters
  Update the centroids as the mean of all data points assigned to each cluster
until convergence
End Procedure

Supplementary Information The online version contains supplementary material available at https://doi.org/10.1007/978-3-032-15176-6_5.

O. Caicedo Rendon, *Machine Learning for Network Management*, Textbooks in Telecommunication Engineering, https://doi.org/10.1007/978-3-032-15176-6_5

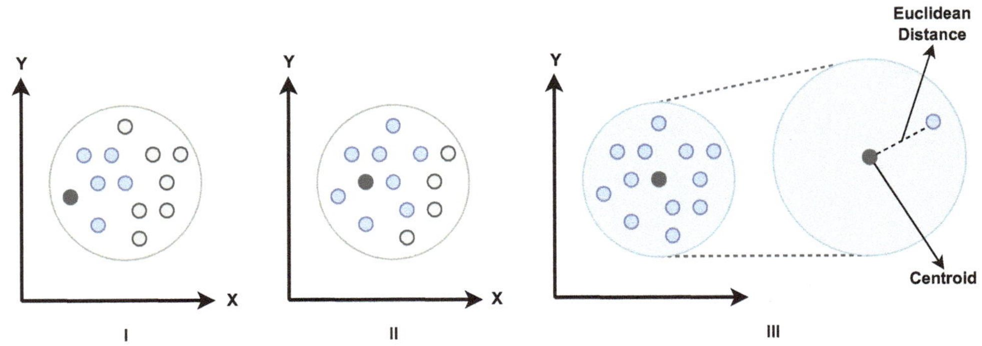

Fig. 5.1 K-means

The K-means algorithm (see Algorithm 1) includes:

- Centroids define the center of the clusters and determine the assignment of data points to clusters during the iterative process of the algorithm. Initially, K centroids are chosen randomly or using a specific initialization method.
- The algorithm assigns each data point to the nearest centroid; each cluster contains data points closer to its centroid than to any other centroid. The assignment of data points to clusters is performed by using the Euclidean distance. This distance is given by Eq. 5.1, where C_i is the set of data points that belong to the i-th cluster. Furthermore, x_j is a data point, μ_i is the centroid of the i-th cluster, and K is the number of clusters.

$$C_i = \{x_j : \|x_j - \mu_i\|^2 \leq \|x_j - \mu_k\|^2 \forall k, 1 \leq k \leq K\} \tag{5.1}$$

- Each centroid is updated to the center of the data points in the corresponding cluster. This update is performed by considering the Eq. 5.2, where μ_i is the updated centroid of the i-th cluster, and $|C_i|$ refers to the number of data points in the i-th cluster.

$$\mu_i = \frac{1}{|C_i|} \sum_{x_j \in C_i} x_j \tag{5.2}$$

- The assignment and updating are repeated until the centroids no longer change significantly, indicating convergence. A predefined number of iterations can also determine convergence.
- K-means is a simple, implementable, and efficient algorithm for large datasets. Furthermore, this algorithm performs well when clusters have a spherical shape. Despite the advantages mentioned above, K-means presents some shortcomings related to its sensitivity to the initial placement of centroids. Furthermore, this algorithm assumes that clusters are of equal size and density and requires that the number of clusters, K, is specified beforehand.

5.1.2 Hierarchical clustering

Hierarchical Clustering (see Fig. 5.2) is a versatile and widely used unsupervised ML algorithm helpful in carrying out clustering tasks, such as network traffic analysis and anomaly detection. This algorithm groups similar data points into clusters based on proximity, creating a tree-like structure called a dendrogram, and builds a hierarchy of clusters through a bottom-up (agglomerative) approach or a top-down (divisive) approach.

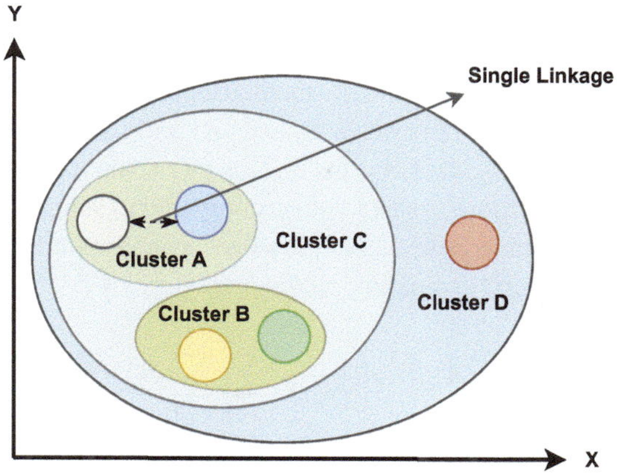

Fig. 5.2 Hierarchical Clustering

Algorithm 2 Hierarchical clustering

Input: Dataset D with data points X, number of clusters K
Output: Hierarchical clustering model H
Procedure:
Initialize each data point as a single cluster
repeat
  Find the pair of clusters that are closest together
  Merge the pair to form a new cluster
  Update the proximity matrix to reflect the new distances
until the desired number of clusters K is reached or all data points are in a single cluster
End Procedure

The Hierarchical Clustering algorithm (see Algorithm 2) includes:

- Each data point is considered a single cluster, forming the initial set of clusters.
- A proximity matrix is created to keep track of the distances between every pair of clusters. The distance between clusters can be measured using various linkage methods, including single linkage, complete linkage, and average linkage.
  - Single linkage measures the distance between two clusters as the shortest distance between any single pair of points in the clusters; see Eq. 5.3 where A and B are clusters, x is a data point in A, and y is a data point in B. This method tends to create long, chain-like clusters and can handle clusters of arbitrary shapes. Nevertheless, it is sensitive to noise and outliers and may result in undesirable chaining effects, where clusters are formed by progressively joining points that are only loosely connected.

$$d_{\text{single}}(A, B) = \min\{d(x, y) : x \in A, y \in B\} \tag{5.3}$$

  - Complete linkage quantifies the distance between two clusters as the maximum distance between any single pair of points in them; see Eq. 5.4 where A and B are clusters, x is a data point in A, and y is a data point in B. This method tends to create compact spherical clusters and is less sensitive to noise and outliers than single linkage. Nonetheless, it may lead to compact clusters that do not accurately represent the natural shape of the data and overemphasize the most significant distances, thereby ignoring the structure within clusters.

$$d_{\text{complete}}(A, B) = \max\{d(x, y) : x \in A, y \in B\} \tag{5.4}$$

- Average linkage is the method in which the distance between two clusters corresponds to the average distance between all pairs of points, one from each cluster. This method operates by following Eq. 5.5, where A and B are clusters, x is a data point in A, y is a data point in B, and $|A|$ and $|B|$ are the number of data points in the corresponding clusters. Average linkage produces clusters that are more balanced in size and shape, less sensitive to outliers compared to single linkage, and more robust to noise compared to single and complete linkage. However, this method can be computationally intensive because it involves averaging all pairwise distances and does not always generate the most intuitive clusters, depending on the dataset.

$$d_{\text{average}}(A, B) = \frac{1}{|A||B|} \sum_{x \in A} \sum_{y \in B} d(x, y) \tag{5.5}$$

- Hierarchical Clustering usually follows an agglomerative approach. Pairs of clusters with the smallest (or maximum or average) distance in the proximity matrix are merged iteratively until a stopping condition is met. This matrix is updated to reflect the distances between the new and the remaining clusters. The process of updating the proximity matrix continues until the desired number of clusters K is achieved or all data points are merged into a single cluster. Notably, it is worth pointing out that this process can also be stopped when a threshold distance is reached.
- Hierarchical Clustering can also follow a divisive approach, where the process starts with all data points in a single cluster. The cluster is then recursively split into smaller clusters. The splitting process involves selecting the cluster with the highest variability or the least cohesion and dividing it into two sub-clusters based on a chosen criterion, such as maximizing the distance between points in different clusters. This process continues until each data point is assigned to its cluster or until a desired number of clusters, K, is reached. The splitting process can also be stopped when a certain threshold distance between clusters is exceeded, ensuring that the clusters formed are sufficiently distinct.

  - Selecting the cluster with the highest variability ensures that the cluster with the most dispersed data points is split first, potentially leading to more homogeneous sub-clusters. In this context, the variability can be quantified using metrics such as variance or standard deviation of the distances between points within the cluster.
  - Selecting the cluster with the least cohesion ensures that the cluster with the most loosely connected data points is split first, which can help in forming more cohesive and meaningful sub-clusters. Cohesion can be measured by the average or total pairwise distance between data points within the cluster.

- Hierarchical Clustering does not require the number of clusters to be specified in advance. Furthermore, it is characterized by producing a dendrogram that provides a visual representation of the clustering process and the relationships between clusters. Moreover, it can handle different types of distance metrics and linkage criteria. Despite its advantages, as mentioned above, it is usually computationally intensive for large datasets and sensitive to noise and outliers.

5.1.3 Density-Based Spatial Clustering of Applications with Noise

DBSCAN (see Fig. 5.3) is a robust and widely used unsupervised ML algorithm that identifies clusters using information about the density of data points, making it particularly effective for discovering clusters of arbitrary shapes and handling noise. This algorithm facilitates clustering tasks in the network security and performance management domain, including anomaly detection and identifying dense regions of network traffic.

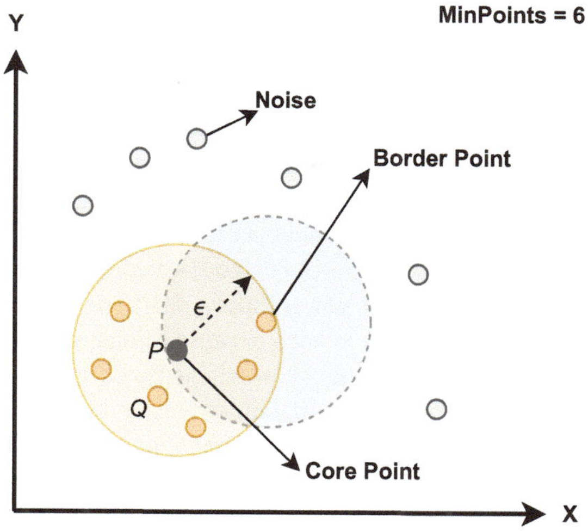

Fig. 5.3 Density-based spatial clustering of applications with noise

Algorithm 3 DBSCAN clustering

Input: Dataset D with data points X, neighborhood radius ϵ, minimum points $MinPts$
Output: DBSCAN clustering model C
Procedure:
Initialize all points as unvisited
for each unvisited point P in D **do**
    Mark P as visited
    Retrieve all points density-reachable from P with respect to ϵ and $MinPts$
    if P is a core point **then**
        Form a cluster C with P and all density-reachable points
    else
        Mark P as noise
    end if
end for
End Procedure

The DBSCAN algorithm (see Algorithm 3) includes:

- Initialization of all data points as unvisited.
- Mark data points as core, border, or noise. A core point has at least $MinPoints$ within a radius ϵ (including itself). A border point has fewer than $MinPoints$ within ϵ but is within the neighborhood of a core point. Noise points are those that are not core points or border points.
- Points directly reachable from a core point within radius ϵ form a cluster. If a point Q is within ϵ distance from a core point P, Q is directly density-reachable from point P. Also, a point Q is density-reachable from point P if there is a chain of points $P_1, P_2, \ldots, P_n$, where $P_1 = P$ and $P_n = Q$, and each P_{i+1} is directly reachable from P_i.
- For each unvisited point P, if P is a core point, a new cluster is formed, including P, and all points are density-reachable from P. If P is not a core point, it is marked as noise unless it is within the neighborhood of a core point.
- DBSCAN does not require knowing the number of clusters in advance. Furthermore, this algorithm can detect clusters of arbitrary shapes and effectively handle noise and outliers. Despite the advantages mentioned above, DBSCAN is computationally intensive for large datasets and is not suitable for clusters with varying densities. Moreover, it requires the selection of values for ϵ and $MinPoints$, which can be a challenging task.

5.1.4 Faults Clustering

K-Means, Hierarchical Clustering, and DBSCAN for Fault Clustering

Framed within fault management, K-means, Hierarchical Clustering, and DBSCAN can be used to categorize historical fault data, helping to diagnose and predict network issues and enabling proactive maintenance, thereby reducing downtime. In a computer network, faults can be broadly categorized into hardware, software, and human; remarkably, each category of fault can impact the network in a different way, and therefore, it requires distinct methods for diagnosis and resolution. Next, for the sake of clarity, these fault categories are briefly discussed. Hardware faults are physical failures or malfunctions of network components. These faults include a router's power supply unit failing or a network cable accidentally cutting, which can disrupt network connectivity, degrade performance, or cause complete network outages. Software faults refer to issues arising from, for instance, firmware bugs, incorrect firewall rule configuration errors, or incompatibilities in app updates for switches, which can lead to performance degradation, security vulnerabilities, or service disruptions. Human faults are errors caused by network practitioners. These faults include actions such as accidental deletion or incorrect setup of routing tables, which can cause traffic to be routed inefficiently and, so, lead to network congestion. According to the methodology discussed in Chap. 3 and beyond the collection, data preprocessing, and feature engineering, for addressing this problem, it is necessary to (i) split the dataset into training and testing sets. (ii) Normalize the features. (iii) Apply the clustering algorithms to learn to categorize different faults. (iv) Evaluate the performance of K-means, Hierarchical Clustering, and DBSCAN models with metrics such as the Silhouette Score, Davies–Bouldin Index (DBI), and Calinski–Harabasz Index (CHI). (v) Once the models are trained and evaluated, they can be deployed to monitor live network traffic and detect faults in real-time. Listing 5.1 shows a Python code snippet to realize the above-described fault classification with K-means, Hierarchical Clustering, and DBSCAN. Readers can use this code as a template for dealing with similar clustering problems.

```
#Import necessary libraries
import numpy as np
import pandas as pd
from sklearn.model_selection import train_test_split
from sklearn.preprocessing import StandardScaler
from sklearn.cluster import KMeans, AgglomerativeClustering, DBSCAN
from sklearn.metrics import silhouette_score, davies_bouldin_score,
    calinski_harabasz_score

#Create a sample dataset
np.random.seed(42)
data = {
    'cpu_utilization': np.random.normal(0, 1, 900),
    'system_uptime': np.random.normal(5, 2, 900),
    'packet_loss_rate': np.random.normal(10, 1, 900),
    'latency': np.random.normal(15, 2, 900),
    'interface_error_rate': np.random.normal(20, 1, 900),
    'memory_utilization': np.random.normal(25, 2, 900),
    'configuration_changes': np.random.normal(30, 1, 900),
    'fault_type': np.random.choice(['hardware', 'software', 'human'],
    900)
}

fault_data = pd.DataFrame(data)

#Drop the 'fault_type' column for clustering
features = fault_data[['cpu_utilization', 'system_uptime', '
    packet_loss_rate', 'latency', 'interface_error_rate', '
    memory_utilization', 'configuration_changes']]

#Normalize the features
scaler = StandardScaler()
features_scaled = scaler.fit_transform(features)

#Split the data into training and testing sets
X_train, X_test = train_test_split(features_scaled, test_size=0.3,
    random_state=42)

#K-means Clustering, k=3, hardware, software, and human faults.
kmeans = KMeans(n_clusters=3, n_init="auto", random_state=42)
kmeans.fit(X_train)
```

```
labels_kmeans = kmeans.predict(X_test)

#Hierarchical Clustering
hierarchical = AgglomerativeClustering(n_clusters=3)
labels_hierarchical = hierarchical.fit_predict(X_test)

#Adjusted DBSCAN parameters
dbscan = DBSCAN(eps=1.5, min_samples=5)
labels_dbscan = dbscan.fit_predict(X_test)

#Evaluate the models
for labels, name in zip([labels_kmeans, labels_hierarchical,
    labels_dbscan], ['K-means', 'Hierarchical', 'DBSCAN']):
    #Check if the algorithm found more than one cluster
    if len(set(labels)) > 1:
        silhouette_avg = silhouette_score(X_test, labels)
        davies_bouldin = davies_bouldin_score(X_test, labels)
        calinski_harabasz = calinski_harabasz_score(X_test, labels)

        #Print evaluation metrics
        print(f"{name} Clustering Model:")
        print(f"Silhouette Score: {silhouette_avg:.2f}")
        print(f"Davies-Bouldin Index: {davies_bouldin:.2f}")
        print(f"Calinski-Harabasz Index: {calinski_harabasz:.2f}")
    else:
        print(f"{name} Clustering Model: No valid clusters found.")

#Deploy the models for real-time fault classification
#This step would involve integrating the trained models into the network
    monitoring system
```

Listing 5.1 Fault clustering

Evaluation Metrics for Clustering Models

The Silhouette Score metric measures the quality of clusters by considering cohesion and separation. In this regard, cohesion refers to the closeness of points within a cluster, while separation indicates the degree of distinctiveness between clusters. Equation 5.6 computes the Silhouette Score $s(i)$ for a data point i, where $a(i)$ is the average distance between i and all other points in the same cluster, representing cohesion, and $b(i)$ is the minimum average distance between i and all points in other clusters to which i does not belong, representing separation. The Silhouette Score values range from -1 to 1. $s(i)$ close to 1 indicates that the data point is well clustered (good cohesion and separation) and close to 0 suggests that the data point is on or very close to the decision boundary between two neighboring clusters. In turn, a Silhouette value close to -1 means that the data point might have been assigned to the wrong cluster.

$$s(i) = \frac{b(i) - a(i)}{\max\{a(i), b(i)\}} \tag{5.6}$$

DBI is a cluster quality metric that evaluates the average similarity of each cluster to the cluster with which it is most similar. DBI considers both the scatter within clusters and the separation between clusters. Equation 5.7 computes DBI, where N is the total number of clusters, σ_i is the average distance between each point in the cluster i and the centroid of the cluster i (scatter within the cluster i), σ_j is the average distance between each point in cluster j and the centroid of the cluster j (scatter within cluster j), $d(c_i, c_j)$ represents the distance between the centroids of clusters i and j (separation between clusters i and j), and $\max_{j \neq i}\left(\frac{\sigma_i+\sigma_j}{d(c_i,c_j)}\right)$ identifies the cluster j that maximizes the ratio of the combined scatter within clusters i and j to the separation between clusters i and j. A DBI close to zero is ideal, indicating distinct and well-defined clusters. A low DBI value indicates good clustering, meaning that the clusters are compact (characterized by low scatter within each cluster) and well-separated (characterized by high separation between clusters). A high DBI value indicates poor clustering, meaning that the clusters have high scatter within them or are not well-separated from each other (clusters overlap).

$$\text{DBI} = \frac{1}{N}\sum_{i=1}^{N}\max_{j \neq i}\left(\frac{\sigma_i + \sigma_j}{d(c_i, c_j)}\right) \tag{5.7}$$

CHI is a metric that measures the quality of clustering by considering the ratio of the sum of dispersion between clusters to the sum of dispersion within a cluster. This metric is given by Eq. 5.8, where k is the total number of clusters, n_i is the number of data points in cluster i, c_i is the centroid of cluster i, c is the centroid of all data points (the overall mean of the dataset), and x represents a data point. Furthermore, $\sum_{i=1}^{k} n_i(c_i - c)^2$ is the sum of the squared distances between each cluster's centroid and the overall centroid, weighted by the number of data points in each cluster (between-cluster dispersion), $\sum_{i=1}^{k} \sum_{x \in c_i} (x - c_i)^2$ is the sum of the squared distances between each data point and its cluster's centroid (within-cluster dispersion), N is the total number of data points, and $\frac{N-k}{k-1}$ is a normalization factor that adjusts for the number of clusters and data points. A high CHI value indicates good clustering, meaning that the clusters are well-separated (with high dispersion between clusters) and compact (with low dispersion within each cluster). A low CHI value suggests poor clustering, meaning that the clusters are not well-separated or not well-compacted.

$$\text{CHI} = \frac{\sum_{i=1}^{k} n_i(c_i - c)^2}{\sum_{i=1}^{k} \sum_{x \in c_i} (x - c_i)^2} \cdot \frac{N-k}{k-1} \tag{5.8}$$

5.1.5 Flows Clustering

K-Means and Hierarchical Clustering for Flows Clustering

K-means and Hierarchical Clustering can categorize network flows into elephants and mice; this categorization can aids in traffic performance management. In a data center network, flows can be categorized into elephants and mice. Each flow category affects the network in a different way and, so, needs distinct routing methods, like ECMP and MiceDCER, to optimize traffic handling. For the sake of clarity, elephant flows refer to large data transfers, such as file downloads or video streams, which can consume significant bandwidth and cause congestion. In turn, mice flows are small, frequent transfers, like web browsing or small file uploads, which generally have minimal impact individually but can collectively affect network performance. According to the methodology discussed in Chap. 3 and beyond the collection, data preprocessing, and feature engineering, for addressing this problem, it is necessary to (i) divide the dataset into training and testing sets. (ii) Normalize the features. (iii) Apply the clustering algorithms to learn to categorize different flows. (iv) Evaluate the performance of the K-means and Hierarchical Clustering models with metrics such as the Silhouette Score, DBI, and CHI. (v) Once the models are trained and evaluated, they can be deployed to monitor live network traffic and detect flows in real-time. Listing 5.2 shows a Python code snippet to realize the above-described flow classification with K-means and Hierarchical Clustering. Note that readers can use this code for addressing similar grouping problems.

```
#Import libraries
import numpy as np
import pandas as pd
from sklearn.model_selection import train_test_split
from sklearn.preprocessing import StandardScaler, LabelEncoder
from sklearn.cluster import KMeans, AgglomerativeClustering
from sklearn.metrics import silhouette_score, davies_bouldin_score,
    calinski_harabasz_score

#Create a sample dataset
data = {
    'start_time': np.random.randint(0, 1000, 500),
    'end_time': np.random.randint(1000, 2000, 500),
    'ip_src': np.random.choice(['192.168.1.1', '192.168.1.2', '
    192.168.1.3'], 500),
    'ip_dst': np.random.choice(['10.0.0.1', '10.0.0.2', '10.0.0.3'], 500)
    ,
    'ip_proto': np.random.choice([6, 17], 500),
    'port_src': np.random.randint(1024, 65535, 500),
    'port_dst': np.random.randint(1024, 65535, 500),
    'pkt_size_1': np.random.randint(40, 1500, 500),
    'pkt_size_2': np.random.randint(40, 1500, 500),
    'pkt_size_3': np.random.randint(40, 1500, 500),
    'pkt_size_4': np.random.randint(40, 1500, 500),
```

```
    'pkt_size_5': np.random.randint(40, 1500, 500),
    'pkt_size_6': np.random.randint(40, 1500, 500),
    'pkt_size_7': np.random.randint(40, 1500, 500),
    'iat_1': np.random.uniform(0, 1, 500),
    'iat_2': np.random.uniform(0, 1, 500),
    'iat_3': np.random.uniform(0, 1, 500),
    'iat_4': np.random.uniform(0, 1, 500),
    'iat_5': np.random.uniform(0, 1, 500),
    'iat_6': np.random.uniform(0, 1, 500),
    'iat_7': np.random.uniform(0, 1, 500),
    'tot_size': np.random.randint(500, 1000000, 500)
}

flow_data = pd.DataFrame(data)

#Convert non-numeric columns to numeric
le_ip_src = LabelEncoder()
le_ip_dst = LabelEncoder()
flow_data['ip_src'] = le_ip_src.fit_transform(flow_data['ip_src'])
flow_data['ip_dst'] = le_ip_dst.fit_transform(flow_data['ip_dst'])

#Drop non-relevant columns for clustering
features = flow_data[
    ['ip_src', 'ip_dst', 'ip_proto', 'port_src', 'port_dst', 'pkt_size_1'
   , 'pkt_size_2', 'pkt_size_3', 'pkt_size_4',
     'pkt_size_5', 'pkt_size_6', 'pkt_size_7', 'iat_1', 'iat_2', 'iat_3',
    'iat_4', 'iat_5', 'iat_6', 'iat_7',
     'tot_size']]

#Normalize the features
scaler = StandardScaler()
features_scaled = scaler.fit_transform(features)

#Split the data into training and testing sets
X_train, X_test = train_test_split(features_scaled, test_size=0.3,
    random_state=42)

#K-means Clustering, k=2 (elephants and mice)
kmeans = KMeans(n_clusters=2, n_init="auto", random_state=42)
kmeans.fit(X_train)
labels_kmeans = kmeans.predict(X_test)

#Hierarchical Clustering
hierarchical = AgglomerativeClustering(n_clusters=2)
labels_hierarchical = hierarchical.fit_predict(X_test)


#Evaluate the models
for labels, name in zip([labels_kmeans, labels_hierarchical], ['K-means',
    'Hierarchical']):
    silhouette_avg = silhouette_score(X_test, labels)
    davies_bouldin = davies_bouldin_score(X_test, labels)
    calinski_harabasz = calinski_harabasz_score(X_test, labels)

    # Print evaluation metrics
    print(f"{name} Clustering Model:")
    print(f"Silhouette Score: {silhouette_avg:.2f}")
    print(f"Davies-Bouldin Index: {davies_bouldin:.2f}")
    print(f"Calinski-Harabasz Index: {calinski_harabasz:.2f}")

# Deploy the models for real-time flow grouping
# This step would involve integrating the trained models into the network
    monitoring system
```

Listing 5.2 Flows grouping

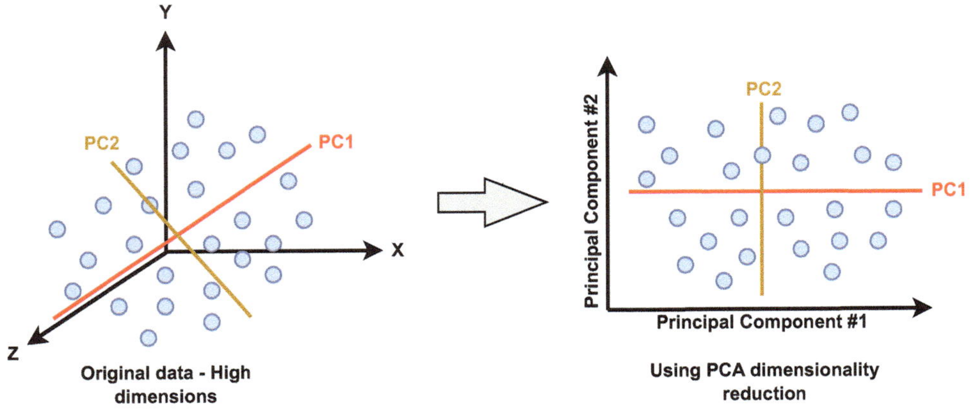

Fig. 5.4 Principal component analysis

5.2 Dimensionality Reduction

With dimensionality reduction algorithms such as PCA, t-SNE, and ICA, network practitioners can, for instance, analyze traffic data and reduce the number of features to the most significant ones, enabling easier detection of unusual traffic patterns that might indicate security breaches. Other problems that can be addressed by the algorithms mentioned include separating overlapping network signals, such as distinguishing between voice and data traffic in a converged network to facilitate QoS management, and visualizing high-dimensional network data in a low-dimensional space, making it easier to understand complex relationships within the data. Moreover, these algorithms can help reduce the dimensions of user behavior data and segment users into categories based on their internet usage patterns for targeted marketing campaigns. Remarkably, after applying PCA or t-SNE, clustering algorithms can be used on the reduced data to categorize traffic types, helping in better bandwidth management and QoS provisioning. Overall, PCA, t-SNE, and ICA algorithms help simplify complex data, making it easier to analyze, visualize, and make informed decisions to enhance network performance and security.

5.2.1 Principal Component Analysis

PCA (see Fig. 5.4) is a widely used unsupervised ML algorithm that helps to reduce dimensionality tasks. In the network management domain, it has helped analyze and visualize traffic to identify critical patterns, relevant features, and trends, facilitate anomaly identification, and optimize resource allocation. This algorithm transforms the original dataset into a new set of orthogonal features, known as principal components, which capture the maximum variance in the data.

Algorithm 4 PCA

Input: Dataset D with data points X, number of principal components k
Output: Principal components $P_1, P_2, \ldots, P_k$
Procedure:
Center the data by subtracting the mean of each feature
Compute the covariance matrix of the centered data
Calculate the eigenvalues and eigenvectors of the covariance matrix
Sort the eigenvalues and their corresponding eigenvectors in descending order
Select the top k eigenvectors to form the principal components
Transform the original data into the new feature space defined by the principal components
End Procedure

The PCA algorithm (see Algorithm 4) includes:

- Center the data by subtracting the mean of each feature to ensure that the dataset has a mean of zero, as shown in Eq. 5.9. Where $\bar{X}$ is the mean of the original matrix X and X_c represents the centered data matrix.

$$X_c = X - \bar{X} \tag{5.9}$$

- Capture the relationships between the features by computing the covariance matrix Σ (with n samples and p features) of the centered data, see Eq. 5.10. Where each element σ_{ij} represents the covariance between the i-th and j-th features, X_c^T is the transpose of the centered data matrix, and $\frac{1}{n-1}$ is used to normalize the covariance computation.

$$\Sigma = \frac{1}{n-1} X_c^T X_c \tag{5.10}$$

- Determine the principal components by computing the covariance matrices of eigenvalues and eigenvectors. Eigenvalues (λ) refer to the amount of variance captured by each principal component, while eigenvectors (v) represent the direction of the principal components (see Eq. 5.11).

$$\Sigma v = \lambda v \tag{5.11}$$

- Sort the eigenvalues and eigenvectors in descendingly order to identify the principal components that capture the most variance.
- Select the top k eigenvectors to form the principal components. Here, k is the number of dimensions for the reduced feature space.
- Transform the original data into the new feature space defined by the principal components, reducing the data's dimensionality while retaining the most essential information. The transformed data matrix Z is given by Eq. 5.12, where $V_k = [v_1, v_2, \ldots, v_k]$ is the top k eigenvectors and X_c represents the centered data matrix.

$$Z = X_c V_k \tag{5.12}$$

- PCA is an effective algorithm that reduces the complexity of high-dimensional data, improves computational efficiency, and facilitates the visualization and interpretation of data. Despite the advantages mentioned, PCA assumes linear relationships between variables and may perform poorly with highly nonlinear data. In addition, this algorithm often requires normalization or standardization because it is sensitive to the scaling of the features.

5.2.2 t-Distributed Stochastic Neighbor Embedding

t-SNE is a robust unsupervised ML algorithm widely used for performing effective dimensionality reduction while preserving local data structure. Framed within the network management areas, this algorithm can help reduce the dimensionality of network traffic data and visualize it, facilitating the identification of clusters of similar traffic patterns and outliers that may indicate potential security threats or intrusions.

Algorithm 5 t-SNE

Input: Dataset D with data points X, number of dimensions d
Output: Low-dimensional representation Y of the data
Procedure:
Compute pairwise similarities P_{ij} in the high-dimensional space
Initialize Y randomly in the low-dimensional space
repeat
  Compute pairwise similarities Q_{ij} in the low-dimensional space
  Minimize the Kullback-Leibler divergence between P and Q
  Update Y using gradient descent
until convergence
End Procedure

The t-SNE algorithm (see Algorithm 5) includes:

- Compute pairwise similarities P_{ij} in the high-dimensional space. These similarities represent the probability that the data points i and j are neighbors. Equations 5.13 and 5.14 define similarities, where $p_{j|i}$ represents the conditional probability that point i would pick point j as its neighbor under a Gaussian distribution centered at i, $\|x_i - x_j\|$ represents the Euclidean distance between i and j in the high-dimensional space, σ_i is the bandwidth of the Gaussian kernel for point i, and N is the total number of data points. Note that, first, the pairwise similarity P_{ij} is symmetrized by averaging $p_{j|i}$ and $p_{i|j}$, and normalized by $2N$ to ensure the similarities sum to 1. Second, σ_i is adjusted iteratively for each data point i to match a set up perplexity value $\text{Perp}(P_i) = 2^{H(P_i)}$, where $H(P_i) = -\sum_j p_{j|i} \log_2 p_{j|i}$ is the Shannon entropy of the conditional probability distribution P_i. The perplexity controls the balance between local and global aspects of the data when mapping high-dimensional data to a lower-dimensional space. A small perplexity value might result in capturing fine-grained structures but might miss broader clusters. A large perplexity value might capture broad clusters but could miss finer details.

$$P_{ij} = \frac{p_{j|i} + p_{i|j}}{2N} \tag{5.13}$$

$$p_{j|i} = \frac{\exp\left(-\frac{\|x_i - x_j\|^2}{2\sigma_i^2}\right)}{\sum_{k \neq i} \exp\left(-\frac{\|x_i - x_k\|^2}{2\sigma_i^2}\right)} \tag{5.14}$$

- Initialize Y randomly in a low-dimensional space (typically with dimensions $d = 2$ or $d = 3$ for visualization purposes).
- Compute pairwise similarities Q_{ij} in the low-dimensional space, representing the probability that the low-dimensional points i and j are neighbors.
- Minimize the Kullback–Leibler (KL) divergence between the high-dimensional and low-dimensional similarity distributions P and Q; see Eq. 5.15 where P_{ij} is the similarity in the high-dimensional space and Q_{ij} is the similarity in the low-dimensional space.

$$KL(P\|Q) = \sum_{i \neq j} P_{ij} \log \frac{P_{ij}}{Q_{ij}} \tag{5.15}$$

- Update Y using gradient descent to minimize the KL divergence, ensuring that similar data points in the high-dimensional space remain close in the low-dimensional space.
- The described process is repeated until convergence, at which point the low-dimensional representation stabilizes.
- t-SNE is computationally intensive, especially for large datasets. Additionally, this algorithm can be highly aware of hyperparameter settings, such as the learning rate and perplexity.

5.2.3 Dimensionality Reduction for Traffic Analysis

PCA for Reducing Dimensionality in Traffic Analysis

PCA can be used to reduce the number of characteristics needed to analyze the network traffic. Following the methodology discussed in Chap. 3 and beyond the collection, data preprocessing, and feature engineering, it is necessary to (i) create a dataset with features including flow duration, the total of forwarding packets, the total backward of packets, the total length of forwarding packets, the total length of backward packets, the maximum length of forwarding packets, and the minimum length of forwarding packets. The target label represents different types of streaming traffic, three classes in this example. (ii) Normalize the features using StandardScaler to improve the models' performance. (iii) Apply PCA to reduce the analysis to 3 principal components. (iv) Evaluate the performance of the reduction by the explained variance ratio. Listing 5.3 shows a Python code snippet to realize the above-described dimensionality reduction with PCA. Readers can use this code as a template to address similar high-dimensionality reduction issues.

```python
#Import necessary libraries
import numpy as np
import pandas as pd
from sklearn.preprocessing import StandardScaler
from sklearn.decomposition import PCA
import matplotlib.pyplot as plt
import seaborn as sns

#Generate a synthetic dataset
np.random.seed(42)
n_samples = 1000

data = {
    'Flow.Duration': np.random.normal(loc=1000, scale=200, size=n_samples),
    'Total.Fwd.Packets': np.random.randint(1, 2000, size=n_samples),
    'Total.Backward.Packets': np.random.randint(1, 2000, size=n_samples),
    'Total.Length.of.Fwd.Packets': np.random.normal(loc=5000, scale=1500, size=n_samples),
    'Total.Length.of.Bwd.Packets': np.random.normal(loc=4000, scale=1200, size=n_samples),
    'Fwd.Packet.Length.Max': np.random.normal(loc=1500, scale=300, size=n_samples),
    'Fwd.Packet.Length.Min': np.random.normal(loc=50, scale=20, size=n_samples),
    'Label': np.random.choice(['Class 1', 'Class 2', 'Class 3'], size=n_samples)
}

synthetic_data = pd.DataFrame(data)

#Select relevant features for PCA
features = synthetic_data[['Flow.Duration', 'Total.Fwd.Packets', 'Total.Backward.Packets',
                           'Total.Length.of.Fwd.Packets', 'Total.Length.of.Bwd.Packets',
                           'Fwd.Packet.Length.Max', 'Fwd.Packet.Length.Min']]

#Normalize the features
scaler = StandardScaler()
features_scaled = scaler.fit_transform(features)

#Apply PCA
pca = PCA(n_components=3)  # Reducing to 3 principal components
principal_components = pca.fit_transform(features_scaled)

#Create a DataFrame with the principal components
pca_df = pd.DataFrame(data=principal_components, columns=['Principal Component 1', 'Principal Component 2', 'Principal Component 3'])

```

```
#Add the target label for visualization
pca_df['Label'] = synthetic_data['Label']

#Plot the PCA result for the first two principal components
plt.figure(figsize=(10, 6))
sns.scatterplot(x='Principal Component 1', y='Principal Component 2', hue
    ='Label', data=pca_df, palette='viridis')
plt.title('PCA of Synthetic Network Traffic Data (First Two Components)')
plt.xlabel('Principal Component 1')
plt.ylabel('Principal Component 2')
plt.legend()
plt.show()

#Plot the explained variance
plt.figure(figsize=(8, 5))
explained_variance = pca.explained_variance_ratio_
plt.bar(range(1, 4), explained_variance, alpha=0.7, align='center', label
    ='Individual Explained Variance')
plt.step(range(1, 4), np.cumsum(explained_variance), where='mid', label='
    Cumulative Explained Variance')
plt.ylabel('Explained Variance Ratio')
plt.xlabel('Principal Components')
plt.title('Explained Variance by Principal Components')
plt.legend(loc='best')
plt.tight_layout()
plt.show()
```

Listing 5.3 Dimensionality reduction with PCA

Explained Variance a Metric for PCA Analysis

The explained variance ratio is a crucial metric in PCA analysis. This metric quantifies the proportion of the total variance in the data that is accounted for a particular principal component. The explained variance, given by Eq. 5.16, provides insight into the importance of each principal component in explaining the variability of the dataset. In this equation, λ_i is the eigenvalue of the i-th principal component, p is the total number of principal components, and $\sum_{j=1}^{p} \lambda_j$ is the sum of all eigenvalues, representing the total variance in the data.

$$\text{Explained Variance Ratio}_i = \frac{\lambda_i}{\sum_{j=1}^{p} \lambda_j} \tag{5.16}$$

$$\text{Cumulative Explained Variance}_k = \sum_{i=1}^{k} \text{Explained Variance Ratio}_i \tag{5.17}$$

The cumulative explained variance is the sum of the explained variance ratios of the principal components up to a specific component. It indicates the percentage of the total variance that is captured by the principal components. This metric is given by Eq. 5.17, where k is the number of principal components considered. If the first few principal components have high explained variance ratios, they capture most of the variability in the data. For instance, by selecting a subset of principal components that together capture a high cumulative explained variance (e.g., 95%), one can reduce the dimensionality of the data while retaining most of its variability. On the other hand, if each principal component has a low explained variance ratio, the variance is more evenly distributed among the components, and more components might be needed to capture a significant portion of the total variance. Note that plotting the explained variance ratio and cumulative explained variance helps visualize the PCA's effectiveness in deciding how many components to retain.

5.3 Outlier Detectors

Outlier detection algorithms, such as Isolation Forest and One-Class SVM, are crucial for realizing security and performance in network management. The use of these algorithms involves detecting anomalous network traffic patterns that may indicate potential intrusions or security breaches, as well as identifying abnormal user behavior that could signify compromised accounts or insider threats. Moreover, Isolation Forest and One-Class SVM can be used to identify rare or unexpected network events that require further investigation or action, and to monitor network performance metrics to detect and address unusual deviations from regular values, ensuring optimal network performance and reliability.

5.3.1 Isolation Forest

Isolation Forest is a compelling ensemble of unsupervised ML algorithms that is helpful, for instance, in analyzing historical data to identify periods of abnormal activity that may correlate with past network issues or outages and in detecting deviations in network traffic that could indicate DDoS attacks, allowing timely mitigation efforts. This algorithm isolates observations by randomly selecting a feature from the dataset. Then, it selects a split value between the maximum and minimum values of the selected feature. The number of splits required to isolate a point is equivalent to the path length from the root to the terminating node. When using Isolation Forest for anomaly detection tasks, such as identifying network intrusions, points that require fewer splits are considered anomalies.

Algorithm 6 Isolation forest

Input: Dataset D with features X, number of trees N
Output: Isolation Forest model F
Procedure:
Initialize F as an empty forest
for $i = 1$ to N **do**
  Draw a random subsample D_i from the dataset D
  Initialize the root node with D_i
  Grow an isolation tree T_i on D_i, $T_i \leftarrow$ GROW_TREE(root, 0)
  Add T_i to the forest F
end for
Define the recursive function GROW_TREE:
function GROW_TREE(node, depth)
  if stopping condition is met **then**
    Assign path length to $node$
    return $node$
  else
    Randomly select a feature A
    Randomly select a split value v between the minimum and maximum values of A
    Split the data at $node$ into $left$ and $right$ subsets based on v
    $node.left \leftarrow$ GROW_TREE($left, depth + 1$)
    $node.right \leftarrow$ GROW_TREE($right, depth + 1$)
    return $node$
  end if
end function
Aggregate the path lengths of the N trees to form the anomaly score
End Procedure

The Isolation Forest algorithm (see Algorithm 6) includes:

- Isolating observations by creating random splits in the data. Points that require fewer splits (shorter path lengths) to be isolated are more likely to be anomalies. The isolation process ensures that anomalous data, which are few and different, are isolated quickly compared to normal data points.
- Splitting the dataset considering the selected feature and split value.
- Recursively applying the splitting function ($grow\_tree(node, depth)$) until a stopping condition is met. Accomplishing a predefined maximum number of nodes and achieving a configured minimum number of samples exemplify such conditions.
- Assigning path lengths to leaf nodes and aggregating the path lengths from all trees to form the final anomaly score.
- Isolation Forest is a robust algorithm that efficiently detects anomalous data. As this algorithm focus on isolating anomalies rather than profiling normal data points, it is particularly effective for high-dimensional datasets. Remarkably, it is essential to be aware that randomness in the selection process performed by Isolation Forest can lead to variability in results, and a significant number of trees may be required to achieve stable results.
- Regarding the differences between Isolation Forest and Random Forest, it is notable that Isolation Forest constructs trees by randomly selecting features and split values. The trees' objective is to isolate data points, and they stop growing when a single data point is isolated or a maximum depth is reached. In contrast, Random Forest builds up trees by selecting the best feature and split value according to a criterion that maximizes the homogeneity of the resulting nodes (e.g., Information Gain); trees grow until all leaves are pure (containing only one class) or another stopping condition is met (e.g., maximum depth, minimum samples per leaf). Also, it is worth noting that Isolation Forest produces an anomaly score for each data point; this score is based on the average path length across all trees (shorter path lengths indicate anomalies). In turn, Random Forest outputs a class label (for classification problems) or a numerical value (for regression problems) by aggregating the outputs of the individual trees (e.g., majority voting for classification or averaging for regression).

5.3.2 One-Class Support Vector Machine

One-Class SVM is a compelling unsupervised ML algorithm for detecting outliers in a dataset, which works by creating a model that can distinguish between the majority of the data points (considered normal) and the origin (a reference "normal" data point) so that any data points that fall outside this boundary are considered outliers. This algorithm is particularly effective for accomplishing network management security, including the identification of anomalous network behavior, detection of fraud, recognition of unusual patterns in time-series data, and detection of abnormal patterns in encrypted traffic. When using One-Class SVM for anomaly detection tasks, such as identifying network intrusions, data points outside the learned decision boundary are considered anomalies.

Algorithm 7 One-class SVM

Input: Dataset D with features X
Output: One-Class SVM model M
Procedure:
Choose a kernel function K
Compute the kernel matrix $K(X_i, X_j)$ for all pairs (X_i, X_j)
Solve the quadratic programming problem to find the optimal hyperplane parameters that separate data points from the origin
End Procedure

The One-Class SVM algorithm (see Algorithm 7) includes:

- One-Class SVM works by projecting the data into a high-dimensional feature space using a kernel (e.g., RBF, introduced in Sect. 4.2, and Sigmoid), and finding a hyperplane that best separates the data from the origin. Points that fall outside the decision boundary are considered anomalies. The Sigmoid kernel behaves similarly to the activation function and is helpful in scenarios where the data follows a pattern that resembles DNN outputs. The combination of the dot product and the hyperbolic tangent function allows the modeling of complex, nonlinear relationships between features. Sigmoid kernel is given by Eq. 5.3.2, where x and y are input vectors, $x \cdot y$ represents the dot product of such vectors, α is a scaling parameter to control the slope of the sigmoid function, c is a constant term to shift such a function to the left or right, influencing the decision boundary's position, and tanh is the hyperbolic tangent function.

$$K(x, y) = \tanh(\alpha(x \cdot y) + c) \tag{5.18}$$

- One-Class SVM can handle high-dimensional data and is effective for detecting novel anomalies in datasets where the majority of the data is normal and anomalies are rare. Despite the mentioned advantages offered by One-Class SVM, this algorithm presents various challenges, including the choice of kernel and its parameters (it requires careful tuning to achieve optimal results), which can significantly impact its performance. Additionally, its performance can be affected negatively by the presence of noise and outliers in the training data.
- While traditional SVM and nonlinear SVMs are supervised ML algorithms used for classification and regression tasks, One-Class SVM is an unsupervised ML algorithm focused on detecting outliers and unusual patterns in data. In particular, the basic SVM finds the optimal hyperplane that disjoints data points of different classes in the feature space. Nonlinear SVM extends traditional SVM by using kernel functions to transform the data into a higher-dimensional space, making it possible to find a hyperplane to separate classes that are not linearly separable in the original space. In turn, One-Class SVM works by finding a decision function that best separates the data points from the origin in the feature space, identifying points outside this decision boundary as anomalies.

5.3.3 IoT Device Abnormal Behavior Detection

Isolation Forest for IoT Device Abnormal Behavior Detection

In network security management, Isolation Forest can identify abnormal IoT device behavior by distinguishing between normal and anomalous activities. According to the methodology discussed in Chap. 3, for addressing this problem, it is necessary (i) to collect IoT device data, including features such as transmitted packets (e.g., a compromised IoT device might participate in a botnet attack, sending an abnormally high number of packets), average packet size (e.g., an IoT device that typically sends small packets might start sending large packets if it is compromised and begins exfiltrating data), communication frequency (e.g., a device that usually communicates every few minutes starts sending data every few seconds, indicating potential malicious activity), CPU usage (e.g., a sudden spike in CPU usage might suggest that a device is executing unauthorized processes or malware.), memory usage (e.g., a device exhibiting a memory leak or running additional unauthorized software might show abnormal memory usage patterns), signal strength (e.g., a device that is suddenly moved to a different location might show a significant change in signal strength), and error rate (e.g., a compromised device might exhibit a high error rate due to tampering or because it is overwhelmed with malicious tasks.). (ii) Generate synthetic data (for demonstration purposes) to simulate normal and abnormal IoT device behavior. (iii) Normalize or standardize the features to ensure they are on a similar scale. (iv) Split the dataset into training and testing sets. (v) Fit the Isolation Forest model to learn to identify anomalies. (vi) Evaluate the performance of the trained Isolation Forest model with the anomaly score. (vii) Once the model is trained and evaluated, it can be deployed to monitor live IoT device data and detect anomalies in real-time. Listing 5.4 shows a Python code snippet to realize the above-described outlier detector with Isolation Forest. Readers can use this code as a template for coping with anomaly problems.

```
#Import Libraries
import numpy as np
import pandas as pd
from sklearn.ensemble import IsolationForest
from sklearn.metrics import silhouette_score, davies_bouldin_score,
    calinski_harabasz_score
import matplotlib.pyplot as plt
import seaborn as sns

#Generate a synthetic dataset for demonstration
np.random.seed(42)
#Normal IoT device behavior data
normal_data = np.random.normal(loc=[100, 500, 10, 30, 40, 80, 0.1], scale
    =[20, 50, 2, 5, 10, 5, 0.05], size=(1000, 7))
#Abnormal IoT device behavior data
abnormal_data = np.random.normal(loc=[1000, 1000, 100, 90, 90, 20, 0.9],
    scale=[50, 100, 10, 2, 2, 2, 0.1], size=(50, 7))
data = np.concatenate([normal_data, abnormal_data])
np.random.shuffle(data)

#Fit the Isolation Forest model
iso_forest = IsolationForest(contamination=0.05, random_state=42)
iso_forest.fit(data)
anomaly_scores = iso_forest.decision_function(data)
labels = iso_forest.predict(data)

# Convert labels from {1, -1} to {0, 1}
labels = np.where(labels == 1, 0, 1)

#Plotting anomaly scores distribution
sns.histplot(anomaly_scores, bins=50, kde=True)
plt.title('Anomaly Scores Distribution')
plt.xlabel('Anomaly Score')
plt.ylabel('Frequency')
plt.show()
```

Listing 5.4 Isolation forest—outlier detector

5.3.4 SDN Switch Traffic Anomaly Detection

One-Class SVM for SDN Switch Traffic Anomaly Detection

In network security management, One-Class SVM can identify anomalous SDN switch traffic by distinguishing between normal and abnormal activities. According to the methodology discussed in Chap. 3, for addressing this problem, it is necessary (i) to collect SDN switch traffic data, including features such as packet count (e.g., a sudden decrease in packet count can signal potential issues such as network failures where the packet count drops significantly), flow count (the number of active flows through the switch is critical for detecting anomalies, a spike in flow count might suggest a flood attack, while a drop could indicate a misconfiguration), packet drop rate (high packet drop rates are often associated with deliberate attacks that aim to disrupt network service), average packet size (abnormal average packet sizes may indicate misconfigured devices sending data at irregular sizes), interarrival time (uniform IATs might suggest automated attacks where packets are sent at regular intervals to overwhelm the network), and protocol type (certain attacks might use specific protocols more frequently, or unusual protocol usage patterns might signal malicious activities). (ii) Generate synthetic data (for demonstration purposes) to simulate normal and anomalous SDN switch traffic. (iii) Normalize or standardize the features to ensure they are on a similar scale. (iv) Split the dataset into training and testing sets. (v) Fit the One-Class SVM model to learn to identify anomalies. (vi) Evaluate the performance of the trained One-Class SVM model with anomaly score. (vii) Once the model is trained and evaluated, it can be deployed to monitor live SDN switch traffic and detect anomalies in real-time. Listing 5.5 shows a Python code snippet to realize the above-described outlier detector with One-Class SVM. Readers can use this code as a template for addressing anomaly detection problems.

```
#Import necessary libraries
import numpy as np
import pandas as pd
from sklearn.svm import OneClassSVM
from sklearn.metrics import silhouette_score, davies_bouldin_score,
    calinski_harabasz_score
from sklearn.preprocessing import StandardScaler
from sklearn.decomposition import PCA
import matplotlib.pyplot as plt
import seaborn as sns

#Generate a synthetic dataset for demonstration
np.random.seed(42)
#Normal SDN switch traffic data
normal_data = np.random.normal(loc=[100, 200, 50, 0.01, 500, 0.1, 1],
    scale=[20, 50, 10, 0.005, 100, 0.02, 0.1], size=(1000, 7))
#Anomalous SDN switch traffic data
anomalous_data = np.random.normal(loc=[1000, 1000, 200, 0.5, 1000, 0.5,
    0], scale=[50, 200, 20, 0.05, 200, 0.05, 0.1], size=(50, 7))
data = np.concatenate([normal_data, anomalous_data])
np.random.shuffle(data)

#Normalize the features
scaler = StandardScaler()
data_scaled = scaler.fit_transform(data)

#Fit the One-Class SVM model
ocsvm = OneClassSVM(kernel='rbf', gamma='scale', nu=0.05)
ocsvm.fit(data_scaled)
labels = ocsvm.predict(data_scaled)

#Convert labels from {1, -1} to {0, 1}
labels = np.where(labels == 1, 0, 1)

#Visual inspection using PCA
pca = PCA(n_components=2)
data_2d = pca.fit_transform(data_scaled)
plt.figure(figsize=(10, 6))
plt.scatter(data_2d[:, 0], data_2d[:, 1], c=labels, cmap='coolwarm', s
    =20, edgecolor='k')
plt.title('One-Class SVM Results (PCA-reduced Data)')
plt.xlabel('PCA Component 1')
plt.ylabel('PCA Component 2')
plt.colorbar(label='Anomaly Score')
plt.show()

#Plotting anomaly scores distribution
sns.histplot(labels, bins=50, kde=True)
plt.title('Anomaly Scores Distribution')
plt.xlabel('Anomaly Score')
plt.ylabel('Frequency')
plt.show()
```

Listing 5.5 One-class SVM—outlier detector

5.4 Summary

This chapter initially introduced various unsupervised ML algorithms, ranging from K-means to One-Class SVM. Furthermore, it exemplified the use of several of them for clustering, outlier detection, and dimensionality reduction tasks from a practical perspective (code snippets in Python were presented) and discussed how unsupervised ML supports various management tasks: faults clustering, flows clustering, IoT device abnormal behavior detection, traffic anomaly detection in SDN switches, and dimensionality dataset reduction for traffic analysis.

Homework Problems and Questions

5.1 What is a cluster? What is an outlier?.

5.2 Explain the metrics used to evaluate clustering and outlier models.

5.3 Detect outliers using One-Class SVM and Isolation Forest in the UNI1 dataset available in the book's repository. Compare the models regarding the anomaly score.

5.4 Group flows with K-means and DBSCAN in the UNI1 dataset available in the book's repository. Compare the models regarding Silhouette Score, Davies–Bouldin Index, and Calinski–Harabasz Index.

Solutions of This Chapter

5.1 Problem. What is a cluster? What is an outlier?.
Solution—Question 1. A cluster embraces a group of data points sharing similarities according to specific criteria, such as distance, density, and connectivity. Intracluster similarity refers to data points within the same cluster being more similar to each other than to those in other clusters. Intercluster dissimilarity refers to data points in different clusters being non-similar. Clustering is an unsupervised ML technique powerful for identifying groups within a dataset.

Solution—Question 2. Outliers are data that significantly deviates from the other data. The Outliers can be caused by, for instance, errors, noise, or unexpected events.

5.2 Problem. Explain the metrics used to evaluate clustering and outlier models.

Solution. There are diverse internal metrics to evaluate clustering models, such as Silhouette Score, Davies–Bouldin Index, and Calinski–Harabasz Index. An internal metric evaluates the quality of clusters based on the data itself. Also, there are external metrics like the Adjusted Rand Index (i.e., it measures the similarity between the predicted clusters and the actual labels, ignoring permutations and with chance normalization—higher values indicate better clustering than lower ones) and Normalized Mutual Information (i.e., measures the mutual information between the predicted clusters and the actual labels, normalized by the entropy of the labels and ignoring permutations—higher values indicate better clustering than lower ones) to assess clustering performance. An external metric evaluates the quality of clusters using external labels (ground truth). On the other hand, for outliers, it is possible to use the same metrics employed for classification tasks (i.e., the evaluation focuses on the model's ability to identify outliers correctly), such as precision, recall, F1-score, and MCC.

5.3 Problem. Detect outliers using One-Class SVM and Isolation Forest in the UNI1 dataset available in the book's repository. Compare the models regarding the anomaly score.

Solving Process. Following the methodology discussed in Chap. 3.

- Business and data understanding. The goal is to classify streaming application traffic. Load the UNI1 dataset using the Pandas library.
- Data engineering.
  - Remove the duplicates and empty instances.
  - Categorize features like IP and MAC addresses to get a dataset with numerical values.
  - Normalize the dataset with StandardScaler.
  - Split the normalized data into train and test sets using 80:20.

- ML model engineering.
  - Import OneClassSVM and IsolationForest from Sklearn library.
  - Build the OneClassSVM-based model with RBF kernel.
  - Create the IsolationForest-based models with 100 estimators.
- Quality assurance for ML applications.
  - Evaluate the OneClassSVM-based models using the test regarding anomaly score.
  - Print the evaluation results using the Matplotlib library.
  - Compare the results.

5.4 Problem. Group flows with K-means and DBSCAN in the UNI1 dataset available in the book's repository. Compare the models regarding Silhouette Score, Davies–Bouldin Index, and Calinski–Harabasz Index.

Solving Process. Following the methodology discussed in Chap. 3.

- Business and data understanding. The goal is to group flows. Load the UNI1 dataset using the Pandas library.
- Data engineering.
  - Remove the duplicates and empty instances.
  - Categorize features like IP and MAC addresses to get a dataset with numerical values.
  - Normalize the dataset with StandardScaler.
  - Split the normalized data into train and test sets using 80:20.
- ML model engineering.
  - Build the K-means-based model with $k = 2$ and $k = 4$.
  - Create the DBSCAN-based model with eps = 0.5 and min_samples = 5.
- Quality assurance for ML applications.
  - Evaluate the models using the test set regarding Silhouette Score, Davies–Bouldin Index, and Calinski–Harabasz Indexes imported from the Sklearn library.
  - Print the evaluation results.
  - Discuss the results.

6 Deep Reinforcement Learning for Network Management

6.1 Algorithms—Overview

RL can be represented as a Markov Decision Process (MDP). Recall that an MDP comprises primarily a state space, an action space, and an immediate reward function [1]. In general, RL algorithms aim to learn an optimal policy that maximizes the long-term reward in the environment by considering the states and actions available for each state. These algorithms can be model-based or model-free. A model-based RL algorithm learns the optimal policy by accessing an environment's model, i.e., a function that predicts the state, action, and reward. A model-free RL algorithm learns an optimal value function (off-policy) or a (deterministic or stochastic) policy (on-policy) by experience. The Actor-Critic, State–Action–Reward–State–Action (SARSA), and Proximal Policy Optimization (PPO) algorithms exemplify on-policy RL. Q-learning may be the most popular off-policy RL algorithm.

DRL is a powerful approach in the field of ML that combines RL and DL. As discussed earlier, RL is a powerful ML technique well-suited for addressing decision-making problems that require automated handling based on trial and error. DL, a subset of ML, harnesses the power of DNN to analyze and interpret complex data, showcasing the potential and capability of this technology. In DRL, a DNN is responsible for learning the value functions or policies through interaction with the environment, facilitated by the RL algorithm. Notably, DRL algorithms such as Deep Q-Network (DQN, also known as Deep Q-Learning), Double DQN (DDQN), Deep Q-learning with prioritized experience replay (prioritized DQN), Dueling DQN, and Distributional DQN are at the forefront of this powerful approach.

The subsequent sections delve into important value-based, policy-based, actor-critic-based, and model-based, as well as centralized and distributed RL and DRL algorithms.

6.2 Value-Based

Value-based RL algorithms estimate the value functions used to evaluate the future expected rewards of different states or of pairs of state-action. The core idea is to utilize value functions to determine which actions to take in order to maximize cumulative rewards over time. This section introduces the most popular RL and DRL value-based algorithms, Q-Learning and DQN.

6.2.1 Q-Learning

Overall, Q-learning (see Fig. 6.1) is a helpful RL algorithm for finding the optimal action for execution in an MDP. This algorithm enables an agent to decide how to act optimally in a discrete and finite environment by learning the value of actions in states in a model-free way, i.e., without requiring an environment's model. Q-learning has been widely enforced in various application domains, including games, robotics, and autonomous systems. Its popularity is due to its effectiveness in learning optimal policies and ease of implementation. Remarkably, in network performance management, it can address admission control, load balancing, and resource allocation problems in computer and telecommunication networks.

Supplementary Information The online version contains supplementary material available at https://doi.org/10.1007/978-3-032-15176-6_6.

O. Caicedo Rendon, *Machine Learning for Network Management*, Textbooks in Telecommunication Engineering, https://doi.org/10.1007/978-3-032-15176-6_6

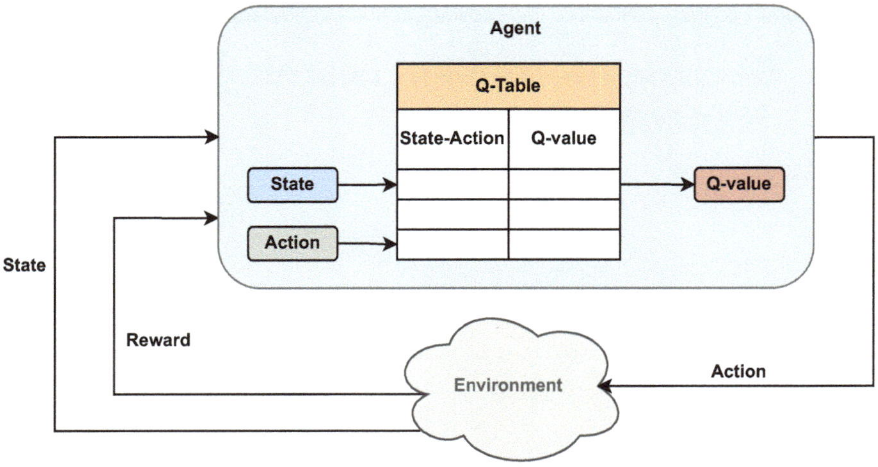

Fig. 6.1 Q-learning

Algorithm 1 Q-learning

Set the learning rate (e.g., $\alpha \leftarrow 0.1$)
Set the discount factor (e.g., $\gamma \leftarrow 0.99$)
Set the number of learning episodes: M
Set the exploration and exploitation parameter: ε
Initialize $Q(s, a)$ for all $s \in$ States, $a \in$ Actions
for episode= 1 to M **do**
  Initialize environment
  Get initial state s
  while not is_terminal_state(s) **do**
    Choose action a from state s using policy (e.g., ϵ-greedy) derived from Q
    Take action a, observe reward r, and next state s'
    Update the current Q-value
    $Q(s, a) \leftarrow Q(s, a) + \alpha \left[r + \gamma \max_{a'} Q(s', a') - Q(s, a)\right]$

    Move to the new state $s \leftarrow s'$
  end while
end for

Q-learning (see Algorithm 1) includes:

- *Agent* is the learner or decision-maker that interacts with the environment to achieve a goal. The agent makes observations, takes actions, and receives rewards to seek such a goal.
- *Environment* refers to the world through which the agent moves. The Environment provides states to the agent and responds to the actions executed by the agent with rewards and new states.
- State Space S is the set of all possible states where the agent can find itself. A state represents the agent's current situation in the environment. Q-learning typically operates with discrete state spaces.
- Action Space A is the set of actions that an agent can perform. For each state, an agent chooses an action on a policy derived from the Q-values. A can be discrete or continuous. In Q-learning, this space is often discrete.
- *Q-values* are updated according to the famous Q-learning formula (see Eq. 6.1), iteratively improving the Policy until it converges to the optimal one that maximizes the cumulative Reward.

$$Q(s, a) \leftarrow Q(s, a) + \alpha[r + \gamma \max_{a'} Q(s', a') - Q(s, a)], \qquad (6.1)$$

where:

- *Q(s, a)* represents the Q-value of taking the action a in the current state s, under a certain policy.
- *Learning Rate* (α) determines to what extent the newly acquired information will override the old information. When α is 0, the agent learns nothing from the actions taken. While α equals 1, the agent considers only the knowledge obtained from the recently carried out actions.
- *Reward* (r) is the reward received after taking action a in state s; it represents the feedback received by an agent from the environment after taking action. A Reward function indicates the immediate gain or loss caused by an action. RL agents aim to maximize the cumulative reward.
- *Discount Factor* (γ) defines the importance of future rewards in the learning process. If γ is 0, the agent only considers current rewards, whereas a discount factor near 1 will encourage it to strive for long-term high rewards.
- The *Target Value* ($r + \gamma \max_{a'} Q(s', a')$) combines the immediate reward r with the best possible future reward $\gamma \max_{a'} Q(s', a')$.
- The Temporal Difference error $[r + \gamma \max_{a'} Q(s', a') - Q(s, a)]$ represents how far off the current estimate is from the target value.

- *Q-table* stores Q-values. This table has rows for each state and columns for each action. The entry in the Q-table for a state-action pair represents the Q-value for that state-action combination. This lookup table is updated with new Q-values as the agent learns through experience.
- *Policy* is the strategy used by the agent to decide the actions to perform at each state. In Q-learning, the Policy is often derived from the Q-values. Greedy Policy is typical in Q-learning. In this Policy, ε-greedy takes an $\varepsilon \in [0, 1]$ as a tunable parameter which allows the agent during the learning process to exploit knowledge with a probability $pr = \varepsilon$ and to explore new ones with a probability $pr = 1 - \varepsilon$ (see Eq. 6.2). A Q-learning agent using the Greedy Policy generates a random value $x \in [0, 1]$ to select the action to perform at a specific state. On the one hand, when $x < \varepsilon$, the agent exploits acquired knowledge (maxQ(s, a′) represents the action a' that maximizes the estimated value for the current state s). On the other hand, if $x > \varepsilon$ the agent explores a random action.

$$a = \begin{cases} \max_{a'} Q(s, a'), & \text{if } x < \varepsilon \\ \text{random action}, & \text{otherwise.} \end{cases} \tag{6.2}$$

- An *Episode* comprises a sequence of steps performed to transition from an initial state to a target state. In turn, a step involves selecting and performing an action, changing the state (i.e., moving from one state to another), and receiving a reward for such an action. The updated value of the Q-function is the underlying reward for executing action A_t in the state S_t, aiming to achieve the optimal reward.

6.2.2 Deep-Q Network

DQN (see Fig. 6.2), a Q-learning extension, is a DRL algorithm that can operate with large spaces of states and actions. This algorithm typically functions with target and online neural networks, which are particularly helpful in determining how to act optimally in a discrete and finite environment by learning the value of actions in states without requiring a model of the environment. Framed within network performance management, DQN can be used to address routing problems in IoT, edge, cloud, softwarized, and mobile communication networks.

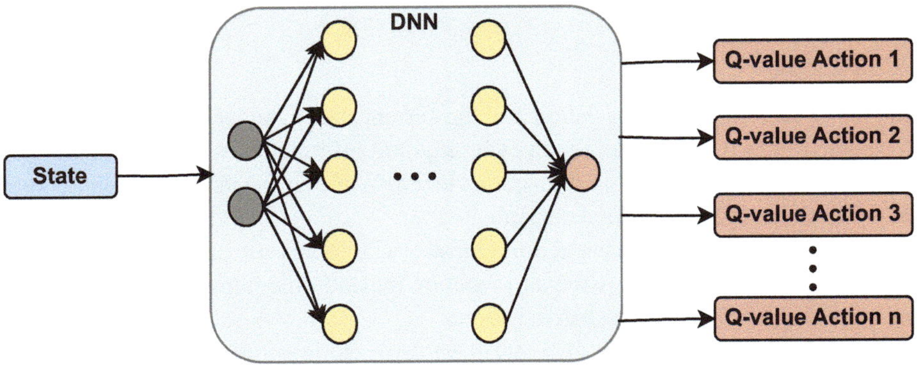

Fig. 6.2 DQN

Algorithm 2 DQN

Set the learning rate (e.g., $\alpha \leftarrow 0.00025$)
Set the discount factor (e.g., $\gamma \leftarrow 0.99$)
Set the number of learning episodes: M
Set the exploration and exploitation parameter: ε_{max} and its decay rate: $decr$
Target update frequency: tup
Initialize replay memory D to capacity rss
Initialize Online DNN with random weights θ
Initialize Target DNN with weights $\theta' = \theta$
for episode= 1 to M **do**
  Initialize environment
  Get state s
  while not is_terminal_state(s) **do**
    Update $\varepsilon = \varepsilon_{max} - (steps \times decr)$
    Select a for s using policy (e.g., ε-greedy)
    Take action a, observe reward r, and next state s'
    Save experience $exp_t = (s, a, s', r)$ into *Replay Memory*
    if $steps > rss$ **then**
      Sample random mini-batch of experiences from *Replay Memory*
      Estimate $Q(s, a : \theta)$ with Online NN
      Estimate $Q(s', a' : \theta')$ with Target NN
      Update Online NN weights minimizing loss (gradient descent and backpropagation)
      $\theta \leftarrow \theta + \alpha \nabla_\theta [r + \gamma \cdot \max_{a'} Q(s', a' : \theta') - Q(s, a : \theta)]^2$
    end if
    if $steps \,\%\, tup == 0$ **then**
      Update θ' of the Target NN and biases with θ of the Online NN
    end if
    Move to the new state $s_t \leftarrow s_{t+1}$
  end while
end for

In addition to the Q-learning elements, DQN (see Algorithm 2) includes:

- DQN can use two DNNs (online and target) to approximate Q-values for each state-action pair.
  - *Online Neural Network* generates the Q-values used to select actions during training. The Online DNN uses gradient descent and backpropagation to adjust its weights, minimizing the loss between target and predicted Q-values at every training step. The loss function is given by Eq. 6.3, where θ and θ' represents the weights of the Online and Target DNNs.

$$\text{Loss}(\theta) = [Q(s', a' : \theta') - Q(s, a : \theta)]^2. \tag{6.3}$$

– *Target Neural Network* generates stable target Q-values for training the Online DNN. The Target DNN synchronizes its weights periodically with the weights of the Online DNN; note that weights are not synchronized at every learning step. The synchronization of weights is a crucial strategy that provides stability to the learning process of DQN, as shown by Eq. 6.4. In this equation, θ and θ' represents the weights of the Online and Target NNs. α, γ, and r are the learning rate, the discount factor and the reward. ∇_θ represents the gradient with respect to the weights θ.

$$\theta \leftarrow \theta + \alpha \nabla_\theta [r + \gamma \cdot \max_{a'} Q(s', a' : \theta') - Q(s, a : \theta)]^2. \tag{6.4}$$

- *Replay Memory* is the buffer D used to store experience tuples. In RL/DRL, an experience tuple, given by Eq. 6.5, captures everything the agent knows at a specific time step. i.e., the state s before the action, the action taken a, the resulting state s', and the immediate reward r obtained after transitioning from s to s'. Remarkably, during training, *mini-batches* of experiences are sampled from the buffer to break the correlation between consecutive experiences, thereby seeking more stable and efficient learning.

$$\text{exp} = (s, a, s', r). \tag{6.5}$$

6.3 Policy-Based

On-policy (D)RL algorithms are designed to optimize the learning policy, selecting actions that maximize expected rewards, making them suitable for environments with continuous action spaces and when stochastic policies are required. Policy-based algorithms are a powerful alternative to value-based ones, especially in complex network management tasks, such as resource allocation in data center networks and route computation in large SDNs.

6.3.1 Proximal Policy Optimization

PPO is a stable and efficient on-policy algorithm that uses a clipped objective function to reduce extensive policy updates (see Fig. 6.3). PPO is not just a theoretical concept; it can be used in real-world applications, such as making real-time decisions on task offloading, resource allocation, and load balancing in edge computing environments and softwarized networks.

PPO (see Algorithm 3) includes:

- *Policy and Value Neural Networks* represent the policy and the value functions. The policy function outputs the probability distribution over actions. The value function estimates the expected return for executing an action from a given state.
- *Clip Range* is the primary innovation PPO offers, which aims to prevent large, potentially destructive updates to the policy. The clipped loss function is given by Eq. 6.6.

$$L^{\text{CLIP}}(\theta) = \hat{\mathbb{E}}_t \left[\min \left(r_t(\theta)\hat{A}_t, \text{clip}(r_t(\theta), 1 - \epsilon, 1 + \epsilon)\hat{A}_t \right) \right]. \tag{6.6}$$

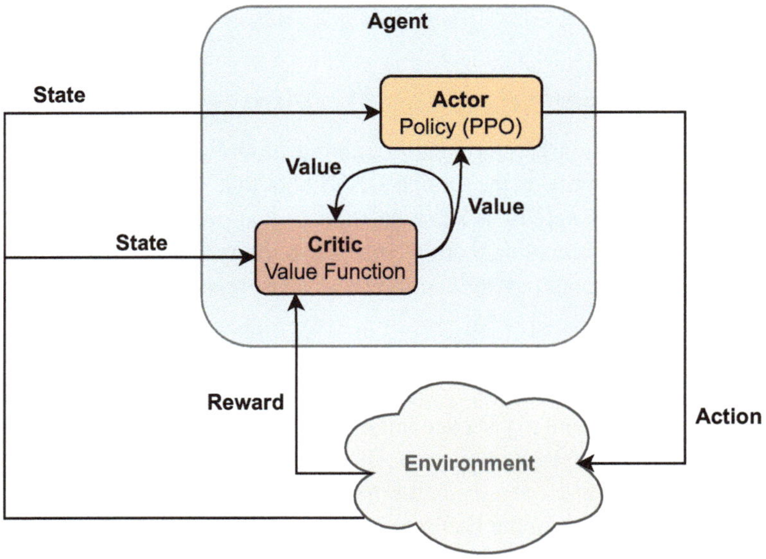

Fig. 6.3 PPO

Algorithm 3 PPO

Initialize Policy NN with parameters θ_0
Initialize Value NN with parameters ϕ_0
for iteration $= 1, 2, \ldots, M$ **do**
  Collect set of trajectories $\mathcal{D}$ by running policy $\pi_{\theta_{\text{old}}}$ in the environment
  for each trajectory **do**
    Compute advantage estimates $\hat{A}_t$
  end for
  for epoch $= 1, 2, \ldots, K$ **do**
    for batch $= 1, 2, \ldots, N$ **do**
      Compute the ratio $r_t(\theta)$
      Compute the clipped loss function $L^{\text{CLIP}}(\theta)$
      Update Policy NN maximizing the surrogate objective

$$\theta \leftarrow \theta + \alpha \nabla_\theta L^{\text{CLIP}}(\theta)$$

      Update Value NN minimizing the loss between the predicted and the actual return

$$\phi \leftarrow \phi - \alpha \nabla_\phi \frac{1}{T} \sum_t \left(V_\phi(s_t) - R_t\right)^2$$

    end for
  end for
  Update $\theta_{\text{old}} \leftarrow \theta$
end for

where:

- $\hat{\mathbb{E}}_t$ indicates that the loss is averaged over a batch of time steps.
- $\min\left(r_t(\theta)\hat{A}_t, \text{clip}(r_t(\theta), 1-\epsilon, 1+\epsilon)\hat{A}_t\right)$ indicates that the loss function always takes the clipped value if the unclipped value result in a large update, which prevents the policy from changing too drastically and stabilizes training.
- $r_t(\theta)\hat{A}_t$ represents the standard policy gradient objective, encouraging the policy to take actions that have higher estimated advantages.
- $r_t(\theta) = \frac{\pi_\theta(a_t|s_t)}{\pi_{\theta_{\text{old}}}(a_t|s_t)}$ indicates how much the policy has changed by comparing the likelihood of the action under the new policy to the old policy. Here, $\pi_\theta(a_t|s_t)$ is the probability of taking action a_t in state s_t under the current policy with parameters θ, $\pi_{\theta_{\text{old}}}(a_t|s_t)$ represents the probability of taking action a_t in state s_t under the old policy before the update.
- $\hat{A}_t$ (Advantage Estimation) serves to guide the direction and magnitude of policy updates since it measures how much better or worse the action a_t is compared to the average action in state s_t. For each trajectory, the advantages estimates $\hat{A}_t$ are computed to reduce the variance of the gradient estimate.
- $\text{clip}(r_t(\theta), 1-\epsilon, 1+\epsilon)$ restricts $r_t(\theta)$ to the range $[1-\epsilon, 1+\epsilon)]$ to prevent the policy ratio from deviating too much from 1, ensuring that policy updates are not too large. Here, ϵ must be a small positive number.

- The *Surrogate Objective* is designed to prevent the policy from changing significantly during each update, thereby ensuring stability. Over several epochs and batches, the Policy NN is updated by maximizing the clipped surrogate objective (see Eq. 6.7) and the value function is updated by minimizing the loss between the predicted value and the actual return (see Eq. 6.8). After the policy and value functions have been updated, the old policy parameters are replaced with the new policy parameters.

$$\theta \leftarrow \theta + \alpha \nabla_\theta L^{\text{CLIP}}(\theta). \tag{6.7}$$

In the Eq. 6.7, θ represents the parameters of the value function V_θ to optimize, α is the learning rate, and ∇_θ denotes the gradient regarding θ and indicates the direction and magnitude of change needed to minimize the loss function computed with Eq. 6.8. Where, $\sum_t$ is the summation over all time steps t, T is the total number of time steps, $V_\phi(s_t)$ is the predicted value in s_t, R_t correspond to the actual return observed at time step t, and $\left(V_\phi(s_t) - R_t\right)^2$ represents the error for each time step computed as the squared difference between the predicted value and the actual return.

$$\frac{1}{T}\sum_t \left(V_\phi(s_t) - R_t\right)^2. \tag{6.8}$$

- A set of *trajectories* $\mathcal{D}$ (known in Q-learning as episodes) refers to a sequence of states, actions, rewards, and other relevant information collected while the agent runs a policy (e.g., $\pi_{\theta_{\text{old}}}$) in the environment. *Mini-batch* processing helps to efficiently utilize collected data and accelerate the learning process during policy optimization. In this regard, an epoch refers to a single pass through the entire batch of collected trajectories (or experience data) for updating the policy and value functions.

6.3.2 Streaming-Proximal Policy Optimization

Still in their early stages in the networking domain, their continuous learning ability, adaptability without forgetting previous knowledge, and faster forecasting than traditional DRL with minibatches make Stream DRL invaluable in driving dynamic environments, such as performance management in 5G and 6G networks. Resource handling, framed within performance management, in 5G and 6G networks, is a complex process that faces various crucial issues. The first is to carefully consider all the network resources, namely RAN, Edge, Transport, and Core. Most existing resource management network slicing solutions are RAN-centered, disregarding one or more core or edge resources needed to model and build slices. The second issue relates to the consideration of admission, allocation, scheduling, and orchestration processes when managing network slice resources. A third issue is the need to operate in highly dynamic and changing environments.

Streaming PPO (S-PPO) is a prime example of a stream-based DRL algorithm. Nevertheless, enforcing S-PPO in the functional areas of network management faces several challenges, including scalability since handling real-time constraints in a large number of nodes and connections requires significant computational resources as well as stability and convergence related to network conditions can change rapidly due to varying traffic patterns, making it difficult for the algorithm to converge to an optimal policy). Other challenges of S-PPO are complexity since the implementation of this algorithm in an existing network management system can be hard, requiring significant changes and coordination with legacy systems, and interpretability and transparency, as understanding decisions can be challenging to interpret and understand the decisions made by the S-PPO agent, making it difficult to troubleshoot and optimize. Moreover, achieving S-PPO explainability is difficult; remarkably, providing clear explanations for the learned policies and their actions is essential for gaining trust and ensuring accountability in network management. Addressing the challenges raised requires careful design, rigorous testing, and continuous monitoring to ensure S-PPO performs effectively and reliably in dynamic network environments.

Algorithm 4 Streaming-PPO

Initialize policy parameters θ and value function parameters ϕ
Initialize the learning rate α, clipping parameter ϵ, discount factor γ, and GAE parameter λ
Initialize buffer $\mathcal{B}$ to store trajectories
for each iteration **do**
  for each actor **do**
    Collect trajectory (s_t, a_t, r_t, s_{t+1}) using policy π_θ
    Store the trajectory in buffer $\mathcal{B}$
    if buffer $\mathcal{B}$ is full **then**
      Compute advantage estimation $\hat{A}_t$
      Compute returns R_t
      Optimize surrogate loss $\mathcal{L}^{CLIP}(\theta)$
      Optimize value function by minimizing the loss

$$\mathcal{L}^{VF}(\phi) = \hat{\mathbb{E}}_t\left[(V_\phi(s_t) - R_t)^2\right]$$

      Update policy parameters θ using gradient ascent

$$\theta \leftarrow \theta + \alpha \nabla_\theta \mathcal{L}^{CLIP}(\theta)$$

      Update value function parameters ϕ using gradient descent

$$\phi \leftarrow \phi - \alpha \nabla_\phi \mathcal{L}^{VF}(\phi)$$

      Clear buffer $\mathcal{B}$
    end if
  end for
end for

S-PPO (see Algorithm 4) is a variation of PPO adapted for streaming data, where policies are updated in small, frequent increments rather than big batch updates, ensuring the model remains responsive to new data. The most relevant differences between PPO and S-PPO are discussed below.

- PPO collects trajectories to compute the advantages and uses them to update the policy and value functions; this process involves collecting a batch of experiences, performing multiple epochs of optimization, and then starting a new batch collection. On the other hand, S-PPO continuously collects and stores trajectories in a buffer, allowing it to make more frequent updates based on a smaller, more recent set of experiences. The ability to rapidly adapt to changing environments is an exciting prospect in streaming RL.

- PPO employs a straightforward batch-oriented implementation that could only be optimal for specific environments. S-PPO involves managing a larger complex buffer. It requires careful handling of the buffer and the optimization process, but it can provide benefits in terms of learning speed and adaptability.

6.4 Actor-Critic-Based

The Actor-Critic is a model-free, online, and on-policy RL algorithm that benefits from value-based (critic) and policy-based (actor) approaches. The actor learns a policy to determine the actions the agent should take in various states, aiming to maximize the cumulative reward through continuous improvement guided by feedback from the critic. The critic evaluates the effectiveness of the actions taken using techniques such as temporal difference, thereby minimizing the error in value estimation. On the one hand, if the learned policy is deterministic, the actor selects a specific action. On the other hand, if the policy is stochastic, the actor chooses a probability distribution over possible actions. In training, the actor and critic networks interact with the environment to collect experiences, which are tuples of state, action, reward, and next state; this process is continuous and iterative, with both networks being updated until the policy converges to an optimal strategy that maximizes cumulative rewards.

6.4.1 Asynchronous Advantage Actor-Critic

A2C is an enhanced version of the traditional Actor-Critic algorithm. What sets A2C apart is the introduction of an advantage function that reduces variance in policy updates, as well as its inherent parallelism. Unlike the traditional Actor-Critic, A2C does not rely on sequential interactions with the environment, but it explicitly runs multiple instances of the environment in parallel. This design choice significantly improves sample efficiency and convergence, as the agent can gather more data simultaneously, showcasing the power of A2C's design.

A3C is also an enhanced version of the traditional Actor-Critic algorithm (see Fig. 6.4). As a result, A3C utilizes an actor to learn the policy and a critic to estimate the value and advantage functions, thereby reducing variance in policy updates and stabilizing the learning process. The pivotal distinction between A2C and A3C lies in their update mechanisms. In A2C, all agents gather experiences and update the global learning parameters synchronously. Conversely, in A3C, multiple agents interact with the environment independently and update the global learning parameters asynchronously. Overall, while A2C is well-known for its stability and reduced variance, A3C is characterized by offering faster convergence due to its asynchronous parallelized approach. A3C in network performance management can help balance the load across multiple servers in cloud environments. Another functional area where A3C can be utilized is configuration management, where an A3C agent can dynamically set up network parameters, such as routing protocols, access control policies, and network topologies, based on real-time data, thereby enhancing network efficiency and resilience.

A3C (see Algorithm 5) includes:

- *Multiple agents (a.k.a., threads)* interact with different environment instances simultaneously, learning independently yet contributing to shared global parameters.
- Each agent interacts with its *environment's copy*, allowing parallel exploration and learning. Remember that the environment is the world through which the agent moves; it provides states to the agent and responds to its actions with rewards and new states.
- Ability to handle discrete or continuous state of spaces S and state of actions A by leveraging neural networks to approximate value functions and policies.

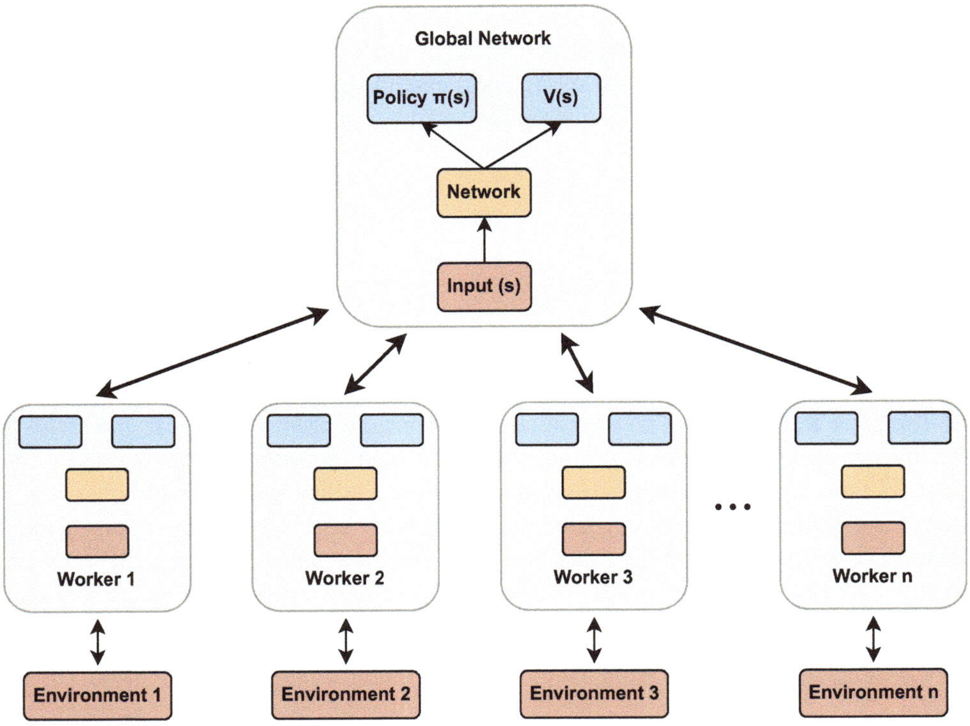

Fig. 6.4 A3C

Algorithm 5 A3C

Initialize global policy parameters θ and value function parameters ϕ
Initialize thread-specific parameters $\theta' \leftarrow \theta$ and $\phi' \leftarrow \phi$
for each actor thread **do**
  repeat
    Reset gradients: $d\theta \leftarrow 0$ and $d\phi \leftarrow 0$
    Synchronize thread-specific parameters: $\theta' \leftarrow \theta$ and $\phi' \leftarrow \phi$
    Get state s_t
    while not terminal state and $t < t_{\max}$ **do**
      Perform action a_t according to policy $\pi(a_t|s_t;\theta')$
      Receive reward r_t and new state s_{t+1}
      $t \leftarrow t+1$
    end while
    $R \leftarrow \begin{cases} 0 & \text{for terminal } s_t \\ V(s_t;\phi') & \text{for non-terminal } s_t \end{cases}$
    for $i \in \{t-1, \dots, 1\}$ **do**
      $R \leftarrow r_i + \gamma R$
      Accumulate gradients wrt θ': $d\theta \leftarrow d\theta + \nabla_{\theta'} \log \pi(a_i|s_i;\theta')(R - V(s_i;\phi'))$
      Accumulate gradients wrt ϕ': $d\phi \leftarrow d\phi + \partial(R - V(s_i;\phi'))^2/\partial\phi'$
    end for
    Perform asynchronous update of θ using $d\theta$
    Perform asynchronous update of ϕ using $d\phi$
  until convergence
end for

- The policy $\pi(a|s;\theta')$ is the agent's strategy to decide its actions at each state parameterized by θ'. The actor-network outputs the probability distribution over actions given the current state, guiding the agent's decisions.
- The *value function* $V(s;\phi)$ estimates the expected return (cumulative future reward) starting from state s and following the policy π; it is represented by a critic network parameterized by ϕ. The critic network helps reduce the variance of the policy gradient updates by providing a baseline for estimating the advantage.

- The term $(R - V(s_i; \phi'))$ represents the advantage, which tells us how much better the action a_i was compared to the expected value. In this regard, R is a reward, and $V(s_i; \phi')$ is the value function that estimates the expected reward for state s_i given parameters ϕ'.
- *Asynchronous updates*. Each agent independently computes gradients based on its interactions with the environment. Those gradients are applied asynchronously to update the global parameters θ and ϕ to help stabilize and speed up the learning process.

  - Equation 6.9 shows how accumulate gradients are computed with respect to θ' to improve the policy. In this equation, $d\theta \leftarrow d\theta$ represents the update of current accumulated gradients with respect to θ' and $\nabla_{\theta'} \log \pi(a_i|s_i; \theta')$ $(R - V(s_i; \phi'))$ accumulates the gradient of the log policy function, weighted by the advantage.

$$\mathrm{d}\theta \leftarrow \mathrm{d}\theta + \nabla_{\theta'} \log \pi(a_i|s_i; \theta')(R - V(s_i; \phi')). \tag{6.9}$$

  - Equation 6.10 shows how accumulate gradients are computed with respect to ϕ' to improve the value function. Where $\mathrm{d}\phi \leftarrow \mathrm{d}\phi$ represents the updating of current accumulated gradients with respect to ϕ' and $\partial(R - V(s_i; \phi'))^2/\partial\phi'$ is the gradient that indicates how much the squared error changes when related to ϕ'.

$$\mathrm{d}\phi \leftarrow \mathrm{d}\phi + \partial(R - V(s_i; \phi'))^2/\partial\phi'. \tag{6.10}$$

- Multiple episodes are run in parallel across different threads, contributing to the agent's global knowledge. Each thread (agent) uses a different learning rate α; remember that α determines the extent to which the new acquired information will override the old information.

6.4.2 Deep Deterministic Policy Gradient

DDPG is an advanced actor-critic and model-free RL algorithm. What sets it apart is its deterministic policy, combined with deep function approximators, a replay buffer, and target networks. With this combination, DDPG learns efficiently and effectively in continuous high-dimensional action spaces. To exemplify, in network performance management, DDPG can be utilized to learn an optimal policy that focuses on adjusting bandwidth distribution to maximize network throughput and minimize latency, ensuring efficient resource utilization. Additionally, this algorithm can help in improving network performance under dynamic network conditions by computing routing paths that prioritize minimizing congestion and delay. Other optimization tasks that DDPG can support are distributing workloads evenly across servers in a data center, seeking to minimize response time and maximize server utilization (no single server is overwhelmed while others are underutilized), prioritizing critical traffic (e.g., Tactile Internet and Remote Surgery) over less critical traffic (e.g., web searching) to ensure a high-quality user experience; and ensuring efficient VNF placement and scaling to maximize resource utilization and maintain service quality in an NFV environment.

The DDPG algorithm (see Algorithm 6) includes:

- A critic network $Q(s, a|\theta^Q)$ to estimate the Q-value for state-action pairs, guiding the actor network's updates.
- An Actor Network $\mu(s|\theta^\mu)$ to determine the action to take given a state, aiming to maximize the expected reward.
- Target networks Q' and μ' to stabilize target values for updates, reducing the probability of divergence in training.
- A replay buffer R to store experiences that allows the algorithm to break the temporal correlations in the sequence of observed states and, as a result, improve learning stability.

Algorithm 6 DDPG

Initialize critic network $Q(s, a|\theta^Q)$ and actor network $\mu(s|\theta^\mu)$ with random weights
Initialize target networks Q' and μ' with weights $\theta^{Q'} \leftarrow \theta^Q, \theta^{\mu'} \leftarrow \theta^\mu$
Initialize replay buffer R
for each episode **do**
  Initialize a random process $\mathcal{N}$ for action exploration
  Receive initial state s_1
  for t = 1 to T **do**
    Select action $a_t = \mu(s_t|\theta^\mu) + \mathcal{N}_t$ according to the current policy and exploration noise
    Execute action a_t and observe reward r_t and new state s_{t+1}
    Store transition (s_t, a_t, r_t, s_{t+1}) in R
    Sample a random minibatch of N transitions (s_i, a_i, r_i, s_{i+1}) from R
    Set $y_i = r_i + \gamma Q'(s_{i+1}, \mu'(s_{i+1}|\theta^{\mu'})|\theta^{Q'})$
    Update critic by minimizing the loss (MSE between predicted and actual return)

$$L = \frac{1}{N}\sum_i (y_i - Q(s_i, a_i|\theta^Q))^2$$

    Update the actor policy using the sampled policy gradient:

$$\nabla_{\theta^\mu} J \approx \frac{1}{N}\sum_i \nabla_a Q(s, a|\theta^Q)|_{s=s_i, a=\mu(s_i)} \nabla_{\theta^\mu}\mu(s|\theta^\mu)|_{s_i}$$

    Update the target networks:

$$\theta^{Q'} \leftarrow \tau\theta^Q + (1-\tau)\theta^{Q'}$$
$$\theta^{\mu'} \leftarrow \tau\theta^\mu + (1-\tau)\theta^{\mu'}$$

  end for
end for

- Minibatch sampling of random experiences obtained from the replay buffer for training the networks.
- Exploration noise $\mathcal{N}$ added to the actor's action for sufficient exploration of the action space.
- Update the actor network by policy gradients.
- Update the critic network by minimizing the MSE between the predicted value and the actual return.
- Smoothly update of target networks to providing stability and reducing variance during the training of the DDPG algorithm, see Eqs. 6.11 and 6.12 where $\theta^{Q'}$ represents the parameters of the target Q-network (critic), θ^Q are the parameters of the main Q-network (critic), $\theta^{\mu'}$ represents the parameters of the target policy network (actor), θ^μ are the parameters of the main policy network (actor), and τ the small constant (between 0 and 1) that determines the soft update rate. Target networks are used to stabilize the training process. They are copies of the main networks but are updated more slowly to provide consistent targets for the Bellman updates, meaning the target networks are not directly set to the main networks' parameters.

$$\theta^{Q'} \leftarrow \tau\theta^Q + (1-\tau)\theta^{Q'} \tag{6.11}$$

$$\theta^{\mu'} \leftarrow \tau\theta^\mu + (1-\tau)\theta^{\mu'}. \tag{6.12}$$

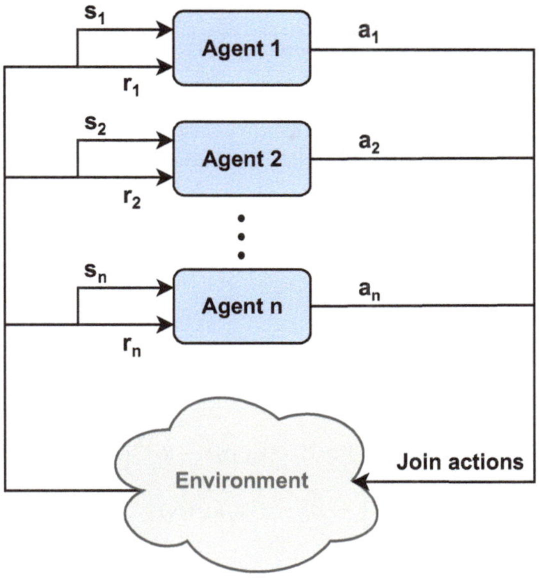

Fig. 6.5 MARL

6.5 Distribution-Based

MARL and FDRL leverage distributed learning; a distributed ML algorithm seeks to improve the learning process regarding response time, memory usage, processing costs, and bandwidth consumption. MARL focuses on decentralized decision-making with potentially centralized coordination, while FDRL emphasizes decentralized data and model training with centralized aggregation. MARL agents learn policies considering the presence of other agents, while FDRL agents collaboratively learn a shared policy. Remarkably, both approaches can handle large and complex environments; MARL scales by adding more agents to the environment, while FDRL scales by leveraging decentralized data and computation.

6.5.1 Multi-agent Reinforcement Learning

As earlier defined, MARL (see Fig. 6.5) is a subset of RL in which multiple agents interact within a shared environment. In this regard, the MRL agents aim to learn optimal behaviors or policies through trial and error, using feedback from the environment. In network performance management, MARL can help, for instance, by optimizing network slices to maximize the profit of providers, achieved by ensuring that each slice receives the necessary resources to meet specific QoS requirements without impacting other slices. This algorithm can also help balance loads dynamically by considering real-time traffic conditions and server capacities, thereby preventing congestion and ensuring efficient resource utilization.

Multi-agent PPO (see Algorithm 7) is an advanced MARL technique. It adapts PPO for use in multi-agent environments, including:

- Agents make decisions and take actions in an environment; remarkably, they need to learn to coordinate their actions to achieve common goals or avoid conflicts. Each agent has a policy that dictates its behavior based on the observations it makes.
- A shared environment where agents interact and receive feedback. The actions of all agents can influence the state of the environment. A state represents the environment at a given time and results often from a combination of the positions or conditions of all agents. Actions are decisions affecting the environment; following an exploration and exploitation strategy, agents must balance exploring new actions to discover effective strategies and exploiting known

Algorithm 7 Multi-agent PPO

Initialize policy parameters θ_i and value function parameters ϕ_i for each agent i
for l **do**each iteration
  for each agent i **do**
    Collect trajectories $\{(s_t, a_t, r_t, s_{t+1})\}$ by running policy π_{θ_i} in the environment
    Compute advantage estimates $\hat{A}_t = R_t - V_{\phi_i}(s_t)$ where R_t is the cumulative reward
    Update policy by maximizing the clipped surrogate objective:

$$L^{CLIP}(\theta_i) = \mathbb{E}_t\left[\min\left(r_t(\theta_i)\hat{A}_t, \text{clip}(r_t(\theta_i), 1-\epsilon, 1+\epsilon)\hat{A}_t\right)\right]$$

where $r_t(\theta_i) = \frac{\pi_{\theta_i}(a_t|s_t)}{\pi_{\theta_i}^{\text{old}}(a_t|s_t)}$

$$\theta_i \leftarrow \theta_i + \alpha \nabla_{\theta_i} L^{CLIP}(\theta_i)$$

    Update value function by minimizing the loss:

$$L_V(\phi_i) = \mathbb{E}_t\left[\left(V_{\phi_i}(s_t) - R_t\right)^2\right]$$

$$\phi_i \leftarrow \phi_i - \alpha \nabla_{\phi_i} L_V(\phi_i)$$

  end for
end for

actions to maximize rewards. Notably, the complexity of MARL-based solutions increases exponentially with the number of agents and the size of the state and action spaces.

- Rewards represent the feedback signals that agents receive after taking an action. The agents are guided by rewards in the pursuit of achieving their objectives.
- Each agent i has its own policy $\pi_{\theta_i}(a_t|s_t)$ that gives the probability of taking action a_t in state s_t. The environment changes as all agents update their policies, making it difficult for any single agent to learn a stable policy. The policy parameters θ_i are updated by maximizing the surrogate objective, as shown in Eq. 6.13.

$$\theta_i \leftarrow \theta_i + \alpha \nabla_{\theta_i} L^{\text{CLIP}}(\theta_i) \tag{6.13}$$

- A clipped surrogate objective to limit the change in the policy. The objective function for agent i is defined in Eq. 6.14.

$$L^{\text{CLIP}}(\theta_i) = \mathbb{E}_t\left[\min\left(r_t(\theta_i)\hat{A}_t, \text{clip}(r_t(\theta_i), 1-\epsilon, 1+\epsilon)\hat{A}_t\right)\right] \tag{6.14}$$

- Value Function estimates the expected cumulative reward an agent can achieve from a given state. The parameters ϕ_i of the value function are updated by minimizing the loss, see Eq. 6.15.

$$\phi_i \leftarrow \phi_i - \alpha \nabla_{\phi_i} L_V(\phi_i). \tag{6.15}$$

- Note that elements in Eqs. 6.13–6.15 were already explained in Sect. 6.3.

6.5.2 Federated Reinforcement Learning

Federated learning, a specific distributed learning approach, enables the construction of accurate models from vast, decentralized, and heterogeneous datasets on resource-constrained devices (e.g., base stations and edge devices). In Federated Learning, multiple nodes or servers collaborate to train a model, but the training data remains decentralized. Remarkably, raw data is not sent to the central server since each device trains the model locally on its own data and only shares the model updates, including gradients or weights. The central server aggregates the updates to improve the global model, thereby enhancing

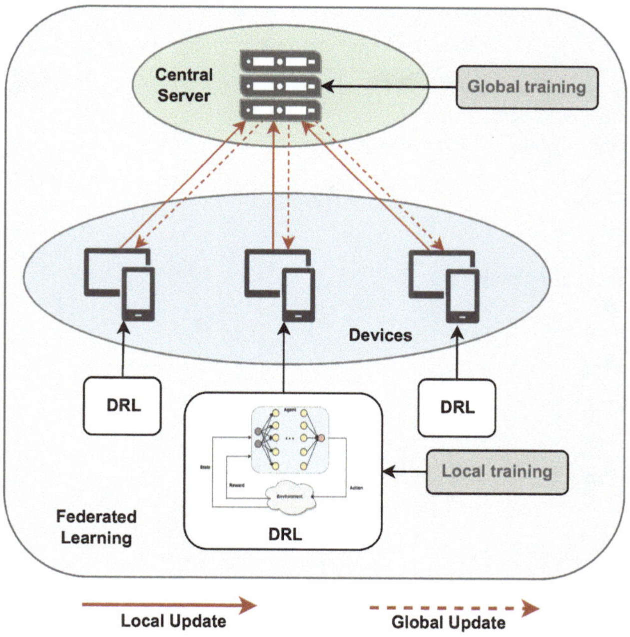

Fig. 6.6 FDRL

data privacy and security, as the raw data never leaves the individual devices. Fundamental advantages of Federated Learning include data privacy (i.e., sensitive data remains on the nodes), bandwidth-friendly (i.e., only model updates are shared, which can be more efficient than transferring large datasets), and scalability (it is possible to leverage the computational power of many devices, enabling the training of large models across distributed networks. Despite its advantages, several challenges must be addressed when using Federated Learning, including ensuring the security of model updates, handling heterogeneous data from various sources, and managing the computational load on devices with differing capabilities.

Federated DRL (FDRL, see Fig. 6.6), a variant of Federated Learning, has been proposed to enhance DRL. In network management, the potential of exploring FDRL in 6G networks is not just pivotal but could be a game-changer in achieving a zero-touch network. Remarkably, most existing routing solutions that utilize RL and DRL algorithms, such as RSIR [2] and DRSIR [3], require more attention to the distributed nature of Internet routing. RSIR capitalizes on the interaction with the environment and the intelligence provided by Q-learning, as well as on the global view and control of the network offered by SDN, to compute and install link-state aware optimal routes in the routing tables of OpenFlow-enabled switches in advance. DRSIR overcomes RSIR scalability issues by calculating path-state aware routes using a centralized DQN with target and online DNNs, replay memory, and mini-batches of experience. A natural evolution for those solutions is creating Federated DRSIR to achieve intelligent and efficient routing in a large SDN. Note that the creation of Federated DRSIR and, overall, the use of FDRL in routing must surpass various challenges related to (i) accomplish a trade-off between processing, memory, bandwidth, and accuracy when computing paths to facilitate their deployment in large networks involving multiple autonomous systems, (ii) build open solutions and experimentation platforms to compare and evaluate FDRL and DRL-based solutions, (iii) obtain routing models that learn distributed and continuously when new data appear; and (iv) safe the coordinator, collaborator nodes, and updates of the shared model.

FDRL (see Algorithm 8) includes:

- Multiple agents collaborate to learn a global policy by periodically synchronizing their local models. In this federated environment, θ^G represents the parameters of the global policy that agents aim to learn, θ^i the local policy parameters, and ϕ^i the parameters for the value function for each agent i.
- Each agent independently performs local training performing a number of local episodes E operating with a learning rate α.

Algorithm 8 FDRL

Initialize global policy parameters θ^G
Initialize local policy parameters θ^i for each agent i to θ^G
Initialize value function parameters ϕ^i for each agent i
Initialize communication rounds K, local episodes E, and learning rate α
for each communication round $k = 1, \ldots, K$ **do**
  for each agent i in parallel **do**
    Initialize local buffer $\mathcal{B}^i$
    for each local episode $e = 1, \ldots, E$ **do**
      Collect trajectory (s_t, a_t, r_t, s_{t+1}) using policy π_{θ^i}
      Store the trajectory in buffer $\mathcal{B}^i$
    end for
    Compute advantage estimation $\hat{A}_t^i$
    Compute returns R_t^i
    Optimize local surrogate loss $\mathcal{L}^{CLIP}(\theta^i)$

$$\mathcal{L}^{CLIP}(\theta^i) = \hat{\mathbb{E}}_t\left[\min\left(\frac{\pi_{\theta^i}(a_t|s_t)}{\pi_{\theta^i_{\text{old}}}(a_t|s_t)}\hat{A}_t^i, \text{clip}\left(\frac{\pi_{\theta^i}(a_t|s_t)}{\pi_{\theta^i_{\text{old}}}(a_t|s_t)}, 1-\epsilon, 1+\epsilon\right)\hat{A}_t^i\right)\right]$$

    Optimize local value function by minimizing the loss

$$\mathcal{L}^{VF}(\phi^i) = \hat{\mathbb{E}}_t\left[(V_{\phi^i}(s_t) - R_t^i)^2\right]$$

    Update local policy parameters θ^i using gradient ascent

$$\theta^i \leftarrow \theta^i + \alpha\nabla_{\theta^i}\mathcal{L}^{CLIP}(\theta^i)$$

    Update local value function parameters ϕ^i using gradient descent

$$\phi^i \leftarrow \phi^i - \alpha\nabla_{\phi^i}\mathcal{L}^{VF}(\phi^i)$$

  end for
  Aggregate local policy parameters to update global policy parameters

$$\theta^G \leftarrow \frac{1}{N}\sum_{i=1}^{N}\theta^i$$

  Broadcast updated global policy parameters θ^G to all agents
  for each agent i in parallel **do**
    Update local policy parameters θ^i to θ^G
  end for
end for

- The agent collects trajectories by interacting with the environment using its local policy π_{θ^i}. The trajectories are stored in a local buffer $\mathcal{B}^i$.
- The agent computes advantages $\hat{A}_t^i$ and returns R_t^i.
- The agent optimizes the local policy by minimizing the surrogate loss $\mathcal{L}^{CLIP}(\theta^i)$ and updates the value function by minimizing the loss $\mathcal{L}^{VF}(\phi^i)$.
- Note that the optimization involves performing gradient ascent for policy parameters and gradient descent for value function parameters.

- The global policy parameters are updated by aggregating the local policy parameters from all agents, typically by averaging them, as shown in Eq. 6.16.

$$\theta^G \leftarrow \frac{1}{N}\sum_{i=1}^{N}\theta^i. \tag{6.16}$$

- The updated global policy parameters θ^G are sent to all agents that, in turn, update their local policy parameters to the new global policy parameters. The update process is repeated K times during the communication rounds.

6.6 Model-Based

As mentioned, DRL can make better decisions than RL in large environments. Despite the benefits offered by DRL, this approach may not be suitable in network management scenarios where the datasets for training DNNs are limited. MBRL, a promising RL approach, leverages a model of the environment to make more informed and efficient decisions, combining the strengths of planning and learning from interaction. An MBRL agent learns a model of the environment to simulate some episodes and plan ahead. In particular, such an agent typically employs a transition model to predict the next state, given the current state and action, and a reward model to predict the reward, given the current state and action. Using the learned models, the agent simulates future trajectories (sequences of states and actions) and evaluates them to find the best sequence of actions that maximizes expected rewards. For action selection, it may employ techniques such as model predictive control.

Recent studies have shown that MBRL methods can achieve similar performance than model-free RL with significantly fewer interactions and limited data, sparking interest in its potential. An MBRL agent can simulate many possible futures without needing to interact with the real environment as much; those simulations can help avoid risky actions in the real world, making it suitable for applications where safety is critical. Furthermore, MBRL enables advanced planning and can adapt to environmental changes more readily than model-free methods. Despite the advantages offered by MBRL, its effectiveness is challenging, requiring careful evaluation of the trade-off between the ease of generating data, the potential bias of model-generated data (an inaccurate model can lead to poor decision-making), and the complexity of building and maintaining a reliable model.

Dyna-Q is an algorithm that integrates model-based and model-free RL. This algorithm utilizes an environment model to simulate experiences alongside real ones to update the Q-values. This unique approach, which emphasizes efficiency, leads to faster convergence and more robust learning. The real-world applications of Dyna-Q can range from establishing optimal transmission paths in environments with malicious jamming to improving routing in wireless mesh networks by optimizing delay and energy efficiency. However, the application of this algorithm in the network management domain still requires in-depth exploration.

Building upon the foundational Q-learning elements, Dyna-Q (see Algorithm 9) introduces the following key features:

- An agent that uses both real experiences from the environment and simulated experiences from its internal environment model to update the Q-values and improve the policy.
- A model $Model(s, a) \leftarrow (r, s')$ that keeps track of the reward and the next state for each state-action pair encountered.
- Planning steps to simulate and learn.
  - The agent randomly selects a previously taken state-action pair $(\hat{s}, \hat{a})$.
  - The agent retrieves the predicted reward $\hat{r}$ and next state $\hat{s}'$ from the model $(\hat{r}, \hat{s}') \leftarrow \text{Model}(\hat{s}, \hat{a})$.
  - The agent uses the simulated experience to update $Q(\hat{s}, \hat{a}) \leftarrow Q(\hat{s}, \hat{a}) + \alpha[\hat{r} + \gamma \max_{a'} Q(\hat{s}', \hat{a}') - Q(\hat{s}, \hat{a})]$.
- Dyna-Q demonstrates its adaptability by using a discrete state space S, which can be seamlessly transformed for continuous spaces using DNNs as function approximators.
- The action space A is usually discrete, allowing the agent to select actions from a finite set of possibilities. This design choice ensures the algorithm's practicality and ease of implementation.

Algorithm 9 Dyna-Q

Initialize Q-values $Q(s, a)$ arbitrarily for all state-action pairs
Initialize the model $Model(s, a)$ to record transitions and rewards
Initialize parameters: learning rate α, discount factor γ, planning steps n
for each episode **do**
  Initialize state s
  repeat
    Choose action a based on $Q(s, a)$ using policy derived from Q
    Take action a, observe reward r and next state s'
    Update Q-values using the observed transition:

$$Q(s, a) \leftarrow Q(s, a) + \alpha[r + \gamma \max_{a'} Q(s', a') - Q(s, a)]$$

    Update the model:

$$Model(s, a) \leftarrow (r, s')$$

    for each planning step from 1 to n **do**
      Randomly select a previously taken state-action pair $(\hat{s}, \hat{a})$
      Retrieve the predicted reward $\hat{r}$ and next state $\hat{s}'$ from the model:

$$(\hat{r}, \hat{s}') \leftarrow Model(\hat{s}, \hat{a})$$

      Update Q-values using the simulated experience:

$$Q(\hat{s}, \hat{a}) \leftarrow Q(\hat{s}, \hat{a}) + \alpha[\hat{r} + \gamma \max_{a'} Q(\hat{s}', \hat{a}') - Q(\hat{s}, \hat{a})]$$

    end for
    $s \leftarrow s'$
  until episode ends
end for

6.7 Methodology for Reinforcement Solutions

Solutions involving RL and DRL algorithms are often complex to construct, evaluate, deploy, and maintain. The following methodology, based on CRISP-ML(Q) and the ML pipeline, helps address this complexity and enables the design of RL-based solutions for the networking area.

The methodology mentioned above comprises the following phases:

- *Business, Problem, and Data Understanding*. This phase is similar to the one defined in the methodology outlined in Sect. 3.3.1. Network practitioners need to answer the following question: What is the problem to solve? Performing intelligent routing in a network that follows the SDN paradigm exemplifies a response to the question raised.
- *Goal Definition*. This phase defines the goal that the RL or DRL algorithm must achieve to address the problem raised in the previous phase. In particular, network practitioners must face the question: What is the goal that the RL or DRL agent needs to accomplish? An example of a response to this question is to select optimal paths by considering bandwidth, delay, and packet loss to decrease network congestion. Path selection has a direct impact on the efficiency and performance of networking systems.
- *Environment Definition*. This phase primarily defines the spaces of states and actions and the reward function.
  - *State Space*. The space S includes all states an RL or DRL agent can experience. Following the example, each state s in S corresponds to an SDN-enabled switch on the Data Plane, and a transition from one state to another corresponds to a link connecting the two corresponding switches. Thus, the exemplified S models the topology of the switches on the Data Plane.

   - *Actions Space*. The space A is the set of all actions an agent of RL or DRL can take. The agent chooses the action a in every learning step, which provides the maximum reward. In the raised example, A is the set of neighboring nodes for each node.
   - *Reward*. The reward function R gives the RL or DRL agent a penalty or award according to the state of the environment reached after it acts. For example, a reward function for intelligent routing can be defined as minimizing delay and packet loss and maximizing throughput.

- *Modeling*. This phase is mainly intended to define the architecture of the RL or DRL algorithm and its corresponding parameters and hyper-parameters.

   - *Agent Architecture*. Network practitioners must initially choose between RL and DRL algorithms to cope with the defined networking problem. RL is suitable for small environments. In turn, DRL is appropriate for large environments. Additionally, practitioners are responsible for parameterizing and hyperparameterizing the chosen algorithm. For instance, in Q-learning, they must define the number of steps and episodes as well as the exploration and exploitation strategy. For DQN, they need to determine, for example, the structure of the Target and Online NNs, including the number of input neurons, hidden layers, output neurons, and the optimization function (e.g., Adam and Stochastic Gradient Descent). Remarkably, a significant issue faced by DRL-based solutions is the hyperparameterization problem.
   - *Agent Training*. This task is similar to the one defined in Sect. 3.3.1. In RL and DRL-based solutions, network practitioners focus on obtaining the optimal set of parameters and hyperparameters that enable the RL or DRL algorithm to achieve the highest reward R while experiencing the shortest convergence time. Following the exemplification of selecting paths, the optimal R must increase throughput while decreasing delay and packet loss.

- *ML Evaluation, Deployment, Monitoring and Maintenance* are similar phases to the methodology outlined in Sect. 3.3.1. Network practitioners must evaluate the practicability of the solution in an actual environment. Pivotal factors of practicability include achieving learning stability when making decisions and minimizing the convergence time of the RL or DRL algorithm as much as possible. In the example mentioned above, good stability and convergence lead to changing paths in a timely manner and only when needed.

6.8 Scaling in Network Functions Virtualization

This section follows the methodology defined in Sect. 6.7 to exemplify the construction of two solutions for scaling a VNF: one based on Q-learning and another based on DQN. The Python code for the two solutions is available on GitHub, which contains the book's examples. The result of conducting the methodology is as follows.

Q-learning for NFV scaling

- *Business, Problem, and Data Understanding.* The problem to be solved is to perform intelligent scaling for a VNF using synthetic data related to the usage of computational resources, specifically in terms of processing and memory.
- *Goal Definition.* The goal is to keep the use of computational resources (CPU and RAM) around 70%.
- *Environment Definition.*

   - The S defined (set of all possible states where the agent can find itself) is made up of two sets. Each set has ten possible discrete states, representing a percentage of the total capacity (ranging from 10 to 100%) in processing and memory. List 6.1 presents a Python code snippet for S.

```
        #Creation of CPU and RAM usage dictionaries
        self.use_CPU = {
            10: (num_CPU * 10) // 100,
```

```
            20: (num_CPU * 20) // 100,
            30: (num_CPU * 30) // 100,
            40: (num_CPU * 40) // 100,
            50: (num_CPU * 50) // 100,
            60: (num_CPU * 60) // 100,
            70: (num_CPU * 70) // 100,
            80: (num_CPU * 80) // 100,
            90: (num_CPU * 90) // 100,
            100: num_CPU
        }
        self.use_RAM = {
            10: (num_RAM * 10) // 100,
            20: (num_RAM * 20) // 100,
            30: (num_RAM * 30) // 100,
            40: (num_RAM * 40) // 100,
            50: (num_RAM * 50) // 100,
            60: (num_RAM * 60) // 100,
            70: (num_RAM * 70) // 100,
            80: (num_RAM * 80) // 100,
            90: (num_RAM * 90) // 100,
            100: num_RAM
        }

```

Listing 6.1 States - Q-learning Agent

- The A (actions the Q-learning agent can take) defined include the following actions: O, which means keeping both computational resources. 1, which means increasing both. 2, which means decreasing both. 3, which means maintaining memory and growing processing. 4, which means maintaining memory and decreasing processing. 5, which means maintain processing and increasing memory. 6, which means maintaining processing and decreasing memory. List 6.2 presents a Python code snippet for A.

```
def step(self, action):
    self.current_step += 1
    #Reaction to q-learning agent actions
    if action == 0: #0. Keep both
        pass
    elif action == 1: #1. Increase both
        self.num_CPU = min(self.num_CPU + self.cpu_10, self.max_CPU)
        self.num_RAM = min(self.num_RAM + self.ram_10, self.max_RAM)
    elif action == 2: #2. Decrease both
        self.num_CPU = max(self.num_CPU - self.cpu_10, 0)
        self.num_RAM = max(self.num_RAM - self.ram_10, 0)
    elif action == 3: #3. Keep RAM and Boost CPU
        self.num_CPU = min(self.num_CPU + self.cpu_10, self.max_CPU)
    elif action == 4: #4. Keep RAM and decrease CPU
        self.num_CPU = max(self.num_CPU - self.cpu_10, 0)
    elif action == 5: #5. Keep CPU and increase RAM
        self.num_RAM = min(self.num_RAM + self.ram_10, self.max_RAM)
    elif action == 6: #6. Keep CPU AND Decrease RAM
        self.num_RAM = max(self.num_RAM - self.ram_10, 0)

```

Listing 6.2 Actions - Q-learning Agent

- List 6.3 presents a Python code snippet for the reward R that is the feedback that the agent receives from the environment after taking an action.

```
#Call reward computation
reward,cpu_reward,ram_reward = self._calculate_reward()

#Update accumulated reward
self.total_reward += reward
self.reward_CPU = cpu_reward
self.reward_RAM = ram_reward
return state, reward, done, {}

#Function for computing reward
```

```
def _calculate_reward(self):
    cpu_reward = 0
    ram_reward = 0

    #Reward related to CPU
    if self.num_CPU <= self.use_CPU[10] or self.num_CPU >= self.
  use_CPU[80]:
        cpu_reward = -50
    elif self.use_CPU[10] < self.num_CPU < self.use_CPU[70]:
        cpu_reward = -1
    elif self.use_CPU[70] < self.num_CPU < self.use_CPU[80]:
        cpu_reward = -1
    elif self.num_CPU == self.use_CPU[70]:
        cpu_reward = 50

    #Reward related to RAM
    if self.num_RAM <= self.use_RAM[10] or self.num_RAM >= self.
  use_RAM[80]:
        ram_reward = -50
    elif self.use_RAM[10] < self.num_RAM < self.use_RAM[70]:
        ram_reward = -1
    elif self.use_RAM[70] < self.num_RAM < self.use_RAM[80]:
        ram_reward = -1
    elif self.num_RAM == self.use_RAM[70]:
    ram_reward = 50

    total_reward = cpu_reward + ram_reward
    return total_reward, cpu_reward, ram_reward

```

Listing 6.3 Reward - Q-learning Agent

- *Modeling and training for Q-learning.* List 6.4 presents a Python code snippet for the Q-learning agent (it implements Algorithm 1) devised and trained to keep the usage of computational resources of a VFN around 70%. This agent uses the spaces of states and actions, the reward, and the epsilon decay policy defined in Lists 6.1–6.3.

```
#Environment initialization
env = VNFEnvironment()

#Q-Learning parameters
alpha = 0.0001  #Learning rate
gamma = 0.95  #Discount factor
epsilon = 1.0  #Maximum scan-exploit parameter
epsilon_min = 0.001 #Minimum scan-exploit parameter
epsilon_decay = 0.9995 #Explore-exploit decay parameter
num_episodes = 15000 #Episodes
max_steps = env.max_step

#Q-table initialization
Q = defaultdict(lambda: np.zeros(env.action_space.n))

#Epsilon-greedy policy to choose an action
def epsilon_greedy_policy(state, epsilon):
    if random.random() < epsilon:
        return env.action_space.sample()  # Exploration
    else:
        return np.argmax(Q[state])  # Exploitation

    #Agent training
    for episode in range(num_episodes):
        state = tuple(env.reset())
        total_reward = 0

        for t in range(max_steps):
            action = epsilon_greedy_policy(state, epsilon)
            next_state, reward, done, _ = env.step(action)
```

```
            next_state = tuple(next_state)

            #Q-table update
            best_next_action = np.argmax(Q[next_state])
            td_target = reward + gamma * Q[next_state][best_next_action]
            td_delta = td_target - Q[state][action]
            Q[state][action] += alpha * td_delta

            total_reward += reward
            state = next_state

            if done:
                break

        #Epsilon decay
        if epsilon > epsilon_min:
            epsilon *= epsilon_decay

        rewards_per_episode.append(total_reward)
        #Record the number of cores used at the end of the episode
        cpu_usage_final_per_episode.append(env.num_CPU)

        if episode % 1 == 0:
            print(f"\nEpisode {episode}/{num_episodes}: Total reward = {total_reward}\nEpisode {episode}/{num_episodes}: Epsilon = {epsilon:.4f}")

```

Listing 6.4 Reward - Q-learning Agent

DQN for NFV scaling

- *Modeling for DQN.* List 6.5 presents a Python code snippet for the DQN agent (it implements Algorithm 2) devised to keep the usage of computational resources of a VFN around 70%. In addition to the spaces of states and actions, the reward, and the epsilon decay policy defined in Lists 6.1–6.3, this agent utilizes an Online DNN, a Target DNN, mini-batches, and a replay memory.

```
#DQNAgent CLASS
class DQNAgent:
  #Constructor
  def __init__(self, state_size, action_size):
    #Pivotal parameters
    self.state_size = state_size
    self.action_size = action_size

    #Memory size
    self.memory = deque(maxlen=5000)

    #Agent parameters
    self.gamma = 0.95
    self.epsilon = 1.0
    self.epsilon_min = 0.01
    self.epsilon_decay = 0.995
    self.learning_rate = 0.001

    #Create models to train target and online DNNs
    self.model = self._build_model()
    self.target_model = self._build_model()
    self.update_target_model()

  #Build the DNN model - Target and Online DNN have the same configuration
  def _build_model(self):
```

```
    #Create the DNN
    model = models.Sequential()
    #Define an input layer according the space of states of the
   environment
    model.add(layers.Input(shape=(self.state_size,)))
    #Define fully connected hidden layers
    model.add(layers.Dense(64, activation='relu'))
    model.add(layers.Dense(64, activation='relu'))
    #Define the output layer according the space of actions of the
   environment <-- Obtain Q-VALUES
    model.add(layers.Dense(self.action_size))
    #Configure the model
    model.compile(optimizer=tf.keras.optimizers.Adam(learning_rate=self
   .learning_rate), loss='mean_squared_error')
    return model

  #Update target model
  def update_target_model(self):
    #Update learned weights for more stable predictions
    self.target_model.set_weights(self.model.get_weights())

  #Experiences function
  def remember(self, state, action, reward, next_state, done):
    #Store experiencies - in memory
    self.memory.append((state, action, reward, next_state, done))

  #Selected actions function
  def act(self, state):
    #Action selection based on epsilon-greedy
    if np.random.rand() <= self.epsilon:
      return random.randrange(self.action_size) #Exploration
    else:
      act_values = self.model.predict(state, verbose=0) #Explotation
      return np.argmax(act_values[0])

  #Replay function
  def replay(self, batch_size):
    #Sample mini-batch
    minibatch = random.sample(self.memory, batch_size)
    #Obtain parameters from the mini-batch
    states = np.vstack([x[0] for x in minibatch])
    actions = np.array([x[1] for x in minibatch])
    rewards = np.array([x[2] for x in minibatch])
    next_states = np.vstack([x[3] for x in minibatch])
    dones = np.array([x[4] for x in minibatch])
    #Predict the targets and next targets
    targets = self.model.predict(states, verbose=0)
    target_next = self.target_model.predict(next_states, verbose=0)
    #Train the DNN with the learned Q-VALUES
    for i in range(batch_size):
      if dones[i]:
        targets[i][actions[i]] = rewards[i]
      else:
        targets[i][actions[i]] = rewards[i] + self.gamma * np.amax(
   target_next[i])
    self.model.fit(states, targets, epochs=1, verbose=0)

  #Update epsilon with decay
  def update_epsilon(self):
    if self.epsilon > self.epsilon_min:
      self.epsilon *= self.epsilon_decay
```

Listing 6.5 Modeling - DQN Agent

- *Training for DQN.* List 6.6 presents a Python code snippet for training the previously designed DQN agent.

```
#Environment - the same defined for Q-learning agent
env = VNFEnvironment()
#Get parameters from the Env
state_size = env.observation_space.shape[0]
action_size = env.action_space.n
#Initialize the DQN agent
agent = DQNAgent(state_size, action_size)
#Define the training parameters
batch_size = 64
episodes = 1000
#List for accumuated reward
rewards_per_episode = []
#Training loop
for e in range(episodes):
  #Reset parameters each peisode
  state = env.reset()
  state = np.reshape(state, [1, state_size])
  total_reward = 0
  done = False
  #Finish episode's loop
  while not done:
    #Select action
    action = agent.act(state)
    #Execute action
    next_state, reward, done, _ = env.step(action)
    #Store reward
    total_reward += reward
    #Update next state
    next_state = np.reshape(next_state, [1, state_size])
    #Store experience
    agent.remember(state, action, reward, next_state, done)
    #Update the current state
    state = next_state
    #Check if the episode finished
    if done:
      #Udate the target model
      agent.update_target_model()
      print(f'\nEpisode {e}/{episodes}: Total reward = {total_reward}\
   nEpisode {e}/{episodes}: Epsilon = {agent.epsilon:.4f}')
      #Store reward per episode
      rewards_per_episode.append(total_reward)
      break
    #Train the DNN of the agent
    if len(agent.memory) > batch_size:
      agent.replay(batch_size)
  #Update epsiolon per episode
  agent.update_epsilon()
```

Listing 6.6 Training - DQN Agent

Metrics for evaluating RL and DRL-based solutions Evaluating RL and DRL agents involves several metrics that can provide insights into the performance, efficiency, and robustness of the learned policies.

- The cumulative reward metric is the total accumulated reward over an episode or a series of episodes. In this regard, a high cumulative reward value indicates a good performance of the RL agent, as it signifies that the agent is effectively achieving its goal(s). Cumulative reward is given by Eq. 6.17, where r_t is the reward at time step t and T is the total number of time steps per episode.

$$R_{\text{cumulative}} = \sum_{t=1}^{T} r_t \tag{6.17}$$

- The reward variance metric measures the variability of the rewards obtained by the agent across different episodes. In this context, a low variance indicates more consistent performance, while a high variance may suggest instability or sensitivity to initial conditions. Reward variance is given by Eq. 6.18, where R_i is the cumulative reward for episode i, μ_R is the mean cumulative reward, and N is the number of episodes.

$$\text{Var}(R) = \frac{1}{N}\sum_{i=1}^{N}(R_i - \mu_R)^2. \tag{6.18}$$

- The episode length is the number of steps taken before reaching a terminal state. Shorter episode lengths can indicate more efficient policies, assuming the agent reaches the goal state. Excessively long episodes may indicate inefficiency or sub-optimal policies.
- The convergence rate measures the speed at which the agent converges to a stable policy. High convergence rates are desirable as they indicate that the agent learns an effective policy quickly, reducing training time and computational resource requirements. The convergence rate is given by Eq. 6.19, where R_e is the cumulative reward for episode e, and E is the total number of episodes.

$$\text{Convergence Rate} = \frac{1}{E}\sum_{e=1}^{E}\left|R_e - R_{e-1}\right|. \tag{6.19}$$

- The policy stability evaluates the degree to which the learned policy remains consistent over time and across different training runs. A stable policy is crucial for reliable and predictable performance, especially in critical applications. The policy stability is given by Eq. 6.20, where $\text{Var}(R)$ is the reward variance and μ_R is the mean cumulative reward.

$$\text{Stability} = 1 - \frac{\text{Var}(R)}{\mu_R}. \tag{6.20}$$

- The balance between exploration and exploitation, given by Eq. 6.21), ensures that the agent can discover optimal policies without getting stuck in local optima.

$$\text{Balance} = \frac{\text{Exploration Steps}}{\text{Total Steps}}. \tag{6.21}$$

- The sample efficiency metric represents the number of training samples required to achieve a certain level of performance. High sample efficiency indicates that the agent learns effectively from limited data, which is crucial in environments where obtaining samples is expensive or time-consuming. The sample efficiency is computed by Eq. 6.22, where R_i is the cumulative reward for episode i and S_i is the number of samples used in episode i.

$$\text{Sample Efficiency} = \frac{1}{N}\sum_{i=1}^{N}\frac{R_i}{S_i}. \tag{6.22}$$

- Generalization refers to the ability of a learned policy to perform well in unseen or slightly different environments compared to the training environment. Good generalization ensures that the RL agent can handle variations and uncertainties in real-world scenarios. Generalization is calculated by Eq. 6.23, where R_j^{test} is the cumulative reward in the test environment j and M is the number of test environments.

$$\text{Generalization} = \frac{1}{M}\sum_{j=1}^{M}R_j^{test}. \tag{6.23}$$

- The safety metric measures the agent's ability to avoid dangerous or undesirable actions that could lead to catastrophic outcomes; it is essential in critical applications, as envisioned in the URLLC 5G use case. Safety is given by Eq. 6.24.

$$\text{Safety} = 1 - \frac{\text{Number of Unsafe Actions}}{\text{Total Actions}}. \tag{6.24}$$

- Computational efficiency measures the computational resources (time, memory, processing power) required to train and execute the RL/DRL algorithm; efficient use of computational resources makes the solution scalable and feasible in practical applications. This metric can be measured by using the computational complexity of the RL or DRL algorithm.
- List 6.7 presents a Python code snippet for computing some of the metrics mentioned above.

```

#Perform imports
import numpy as np

#Assume the rewards from 10 episodes
rewards_per_episode = [20, 25, 30, 28, 22, 27, 29, 31, 24, 26]

#Calculate Cumulative Reward
cumulative_reward = sum(rewards_per_episode)

#Calculate Average Reward per Episode
average_reward = np.mean(rewards_per_episode)

#Calculate Reward Variance
reward_variance = np.var(rewards_per_episode)

#Assume the episode lengths is known
episode_lengths = [50, 45, 60, 55, 48, 52, 49, 60, 50, 53]

# Calculate Average Episode Length
average_episode_length = np.mean(episode_lengths)

# Display the metrics
print(f"Cumulative Reward: {cumulative_reward}")
print(f"Average Reward per Episode: {average_reward}")
print(f"Reward Variance: {reward_variance}")
print(f"Average Episode Length: {average_episode_length}")


```

Listing 6.7 Training - DQN Agent

- Remarkably, evaluating RL and DRL agents involves also measure the impact in the network performance by using metrics such as delay, packet loss, throughput, reliability, and profit.

6.9 Summary

This chapter initially presented how an RL or DRL agent, on its journey of learning and decision-making, executes various actions within an environment to achieve specific objectives. The learning process is guided by positive rewards for desirable outcomes (states) resulting from the actions and negative for undesirable ones. These rewards serve as a guide, helping the agent understand which actions are beneficial and which are not. As the agent accumulates experience, it develops a policy that links environment states to actions. This policy is crucial as it guides the agent's decision-making, aiming to maximize cumulative reward. Later, this chapter also discussed various RL and DRL algorithms. Then, it presented how RL and DRL can support various management tasks. Finally, it introduced a methodology to create, evaluate, deploy, and maintain RL or DRL-based solutions and showed its use by practical examples and their corresponding implementation code.

Homework Problems and Questions

6.1 Explain how RL and DRL centralized algorithms operate.

6.2 Discuss how RL and DRL distributed algorithms function.

6.3 Model RL (Q-learning and SARSA) and DRL (DQN and PPO) agents to handle the number of User Plane Functions needed to support a uRLLC service. Compare the models regarding convergence time, computational resources consumption, and scaling decisions.

6.4 Model RL (Q-learning) and DRL (PPO and A3C) agents to handle the number of Admission Management Functions needed to support an mMTC service. Compare the models regarding convergence time, computational resources consumption, and scaling decisions.

Solutions of This Chapter

7.1 Problem. Explain how RL and DRL centralized algorithms operate.

Solution. RL is an ML field for learning optimal policies through interaction with an environment. Its pivotal components are an agent that makes decisions, an environment representing the external system with which the agent interacts, a states space that represents the possible environment states, an actions space representing a decision taken by the agent that affects the environment, a reward function that offers a scalar feedback signal indicating the immediate benefit of taking action in a state, a policy that represents the strategy used by the agent to select actions based on states and a value function that estimates the expected cumulative reward from a state (or state-action pair) under a given policy. DRL extends RL by DNNs to approximate the value function or policy, which allows DRL agents to handle high-dimensional state and action spaces, which are common in network management. A single agent is responsible for learning and decision-making in centralized RL and DRL. The agent has full access to the state of the environment and computes actions based on its policy.

7.2 Problem. Discuss how RL and DRL distributed algorithms function.

Solution. The pivotal idea in distributed RL and DRL is to parallelize the learning process across multiple agents to collect data faster through parallel exploration, use computational resources more efficiently, and improve scalability for large computer and telecommunication networks. Remarkably, in a distributed setting, each RL/DRL agent interacts with its instance of the environment, collects experiences independently, and acts based on its local policy but periodically synchronizes with the global policy handled by a central server that aggregates updates (synchronous or asynchronously) from all workers and maintains a global policy (or value function).

7.3 Problem. Model RL (Q-learning and SARSA) and DRL (DQN and PPO) agents to handle the number of User Plane Functions needed to support a uRLLC service. Compare the models regarding convergence time, computational resources consumption, and scaling decisions.

Solving process. Following the methodology discussed in this chapter.

- Business and data understanding. The goal is to handle the number of User Plane Functions needed to support a uRLLC service. For this kind of service, it is necessary to collect data about delay, packet loss, and computational resources per UPF.
- Define a discrete states space for Q-learning and SARSA agents, it should consider a UPF identifier, the current packet loss and delay suffered by the service slice, and the available memory and processing per UPF.
- Define a continuous states space for DQN and PPO agents, it should consider a UPF identifier, the current packet loss and delay suffered by the service slice, and the available memory and processing per UPF.
- Define a discrete actions space for Q-learning and SARSA agents, it could maintain, reduce, and increase the number of UPFs.
- Define a continuous actions space for DQN and PPO agents, it could maintain, reduce, and increase the number of UPFs.
- Model the reward function, it should be positive when there are low latency, efficient use of UPFs, and low packet loss rate. On the contrary, this functions should penalty when high latency, high packet loss, and widespread changes in the number of UPFs cause instability.
- Build the environment with Gymnasium—an environment for Q-learning and SARSA and another one for DQN and PPO agents.

- Create and train the Q-learning and SARSA agents with Gym, Numpy, Pandas, the collected data and the corresponding environment.
- Create and train the DQN and PPO agents with with Gym, Numpy, Pandas, the collected data and the corresponding environment.
- Evaluate the Q-learning and SARSA agents regarding convergence time, computational resources consumption, and scaling decisions.
- Evaluate the DQN and PPO agents regarding convergence time, computational resources consumption, and scaling decisions.
- Discuss the results when varying learning rate and actions selection strategy, for instance.

7.4 Problem. Model RL (Q-learning) and DRL (PPO and A3C) agents to handle the number of Admission Management Functions needed to support an mMTC service. Compare the models regarding convergence time, computational resources consumption, and scaling decisions.

Solving process. Following the methodology discussed in this chapter.

- Business and data understanding. The goal is to handle the number of User Plane Functions needed to support an mMTC service. For this kind of service, it is necessary to collect data about admission success rate, admission failure rate, admission latency, and computational resources per AMF.
- Define a discrete states space for the Q-learning agent, it could consider the current number of active AMFs, admission success rate, admission failure rate, average admission latency, and use of computational resources of AMFs.
- Define a continuous states space for PPO and A3C agents, it could consider the current number of active AMFs, admission success rate, admission failure rate, average admission latency, and use of computational resources of AMFs.
- Define a discrete actions space for the Q-learning agent, it could activate, deactivate, and keep the current number of AMFs.
- Define a continuous actions space for PPO and A3C agents, it could maintain, reduce, and increase the number of UPFs.
- Model the reward function, it should be positive when there is a high success rate in device admission, low latency in the admission process, and efficient use of AMFs (AMFs are the optimal number without wasting resources). On the contrary, it should penalty when there is a high admission failure rate, high admission latency, and excess of active AMFs.
- Build the environment with Gymnasium—an environment for Q-learning and another one for PPO and A3C agents.
- Create and train the Q-learning agent with Gym, Numpy, Pandas, the collected data and the corresponding environment.
- Create and train the PPO and A3C agents with with Gym, Numpy, Pandas, the collected data and the corresponding environment.
- Evaluate the Q-learning agent regarding convergence time, computational resources consumption, and scaling decisions.
- Evaluate the PPO and A3C agents regarding convergence time, computational resources consumption, and scaling decisions.
- Discuss the results when varying learning rate and actions selection strategy, for instance.

References

1. T.T. Nguyen, N.D. Nguyen, S. Nahavandi, Deep reinforcement learning for multiagent systems: a review of challenges, solutions, and applications. IEEE Trans. Cybern. **50**(9), 3826–3839 (2020)
2. D.M. Casas-Velasco, O.M.C. Rendon, N.L.S. da Fonseca, Intelligent routing based on reinforcement learning for software-defined networking. IEEE TNSM **18**(1), 870–881 (2020)
3. D.M. Casas-Velasco, O.M. Caicedo Rendon, N.L.S. da Fonseca, DRSIR: a deep reinforcement learning approach for routing in software-defined networking. IEEE Trans. Netw. Service Manage. **19**(4), 4807–4820 (2022). https://doi.org/10.1109/TNSM.2021.3132491

7 Network Management Assisted by Generative Learning

7.1 Autoencoders and Variants

7.1.1 Autoencoders

Autoencoders are a powerful generative AI model designed to minimize the difference between the input data and its reconstruction (see Fig. 7.1). By encoding input data into a lower-dimensional latent space and then reconstructing it, autoencoders can generate new data similar to the original dataset. Remarkably, an autoencoder comprises mainly two parts: the Encoder and the Decoder. The decoder compresses the input data into a lower-dimensional representation, meaning that it maps the input data $\mathbf{x}$ to a latent space $\mathbf{z}$. The decoder reconstructs the input data from the lower-dimensional representation. It maps the latent space $\mathbf{z}$ back to the input space, trying to reproduce $\mathbf{x}$. In diverse application domains, autoencoders have been helpful for tasks such as dimensionality reduction, image denoising, and data compression.

In network-fault management, autoencoders can be used for proactive maintenance by predicting future network failures by identifying patterns from historical network data that precede faults. In configuration management, autoencoders can suggest optimized configurations to enhance network performance and reliability by validating network configurations throughout the learning of standard configurations and identifying deviations. In security and accounting management, autoencoders can detect malware by identifying abnormal traffic patterns that differ from the learned normal behavior and detect unusual accounting record patterns that can indicate fraudulent activities. In performance management, autoencoders can analyze network traffic to detect deviations that might indicate a drop in QoS, leading to a breach of SLAs, and identify performance bottlenecks. In summary, by learning and recognizing patterns in network data, autoencoders can enhance all functional areas of network management.

Algorithm 1 Autoencoder

```
Input: Dataset D with features X
Output: Generative autoencoder model GA
Procedure:
Initialize the network with random weights and biases
for each epoch do
  for each batch of data do
    Forward propagate the input through the encoder to compute the latent representation
    Forward propagate the latent representation through the decoder to reconstruct the input
    Compute the reconstruction loss between the original input and the reconstructed input
    Backpropagate the error to update the weights and biases using gradient descent
  end for
end for
End Procedure
```

Supplementary Information The online version contains supplementary material available at https://doi.org/10.1007/978-3-032-15176-6_7.

O. Caicedo Rendon, *Machine Learning for Network Management*, Textbooks in Telecommunication Engineering, https://doi.org/10.1007/978-3-032-15176-6_7

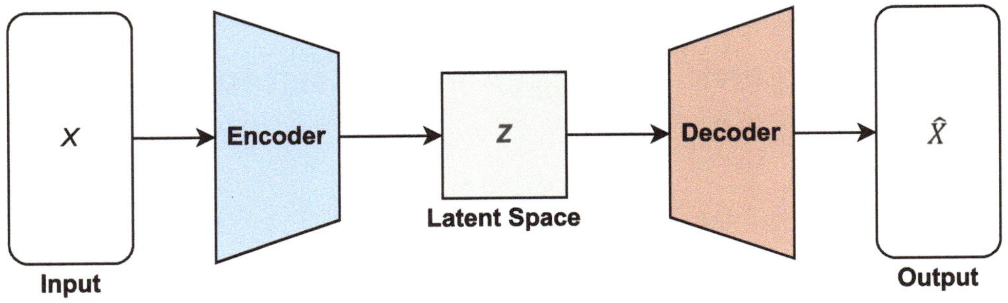

Fig. 7.1 Generative autoencoder architecture

The generative autoencoder training process (see Algorithm 1) includes:

- Initialize the autoencoder with random weights and biases.
- Iteratively improve the model by adjusting the weights and biases through forward propagation, computing the loss, and backpropagation.
  - During forward propagation, the input data passes through the encoder to produce a latent representation and then through the decoder to reconstruct the input. The latent representation (or latent space) is the zipped, lower-dimensional version of the input data produced by the encoder. If x is the input data and f_{encoder} is the encoder function, the latent representation z is given by Eq. 7.1. Remarkably, (i) the dimensionality reduction forces the autoencoder to capture the input data's most critical features (crucial for data reconstruction). (ii) The latent space acts as a bottleneck, ensuring that the DNN discards noise and redundant information and that only the most significant information is passed to the decoder. (iii) The encoder can be a feature extractor for other ML algorithms. (iv) The latent space can be visualized in 2D or 3D using t-SNE or PCA to understand how the autoencoder organizes the data; this visualization often reveals clusters of similar data points, providing insights into the learned features.

$$\mathbf{z} = f_{\text{encoder}}(\mathbf{x}) \tag{7.1}$$

  - Each neuron in the encoder and decoder behaves as follows. (i) It computes a weighted sum of its inputs. (ii) It adds a bias. (iii) It applies an activation function to introduce nonlinearity into the model, enabling the network to learn complex patterns. (iv) It passes the result of the activation function to the next layer.
  - A loss function measures the error between the original input and the reconstructed input, guiding the optimization process. A typical loss function for autoencoders is MSE.
  - During backward propagation, the network computes the gradient of the loss function with respect to each weight and bias using the chain rule. Once these gradients are calculated, an optimization algorithm adjusts the weights and biases to minimize the loss function. Typical optimization algorithms for autoencoders are SGD, RMSprop, and Adam.

- The final model consists of optimized weights and biases that minimize the loss function, providing accurate reconstructions for new input data.
- Autoencoders are versatile and can model complex, nonlinear relationships in the data by adjusting the architecture (including the number of layers and neurons) of the encoder and decoder, as well as hyperparameters. Despite their advantages, autoencoders are challenging to interpret and require significant computational resources for training, especially with large datasets and deep architectures. Moreover, autoencoders are sensitive to hyperparameter tuning, requiring careful selection of learning rates, batch sizes, and architectures for optimal performance. In addition, training deep autoencoders can lead to issues such as vanishing or exploding gradients and overfitting, which can hinder convergence. To address these challenges, techniques like normalization, advanced initialization, and dropout are often employed.

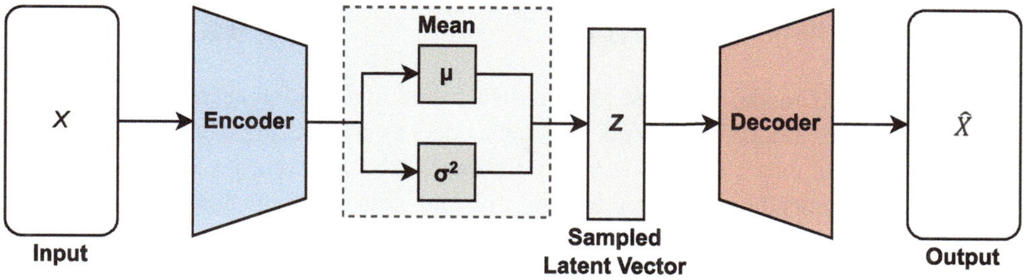

Fig. 7.2 Variational autoencoder architecture

- Dropout is a regularization technique used in DNNs to prevent overfitting and improve generalization; this technique is also characterized by its ease of implementation and computational inexpensiveness. In this vein, in the context of autoencoders, dropout can be applied to both the encoder and decoder parts to enhance the model's robustness and generalization.
- Dropout operates as follows. (i) For each training example, it randomly selects a fraction p of the neurons to drop. (ii) The selected neurons are temporarily removed from the DNN for that particular forward and backward pass. (iii) To maintain the overall scale of the activations, the outputs of the remaining neurons are scaled by $\frac{1}{1-p}$ during training. Without dropout, the activations in the encoder and decoder layers are computed by Eqs. 7.2 and 7.3. Where W_1 and W_2 are the weight matrices, b_1 and b_2 are the bias vectors, and f_{encoder} and f_{decoder} are the activation functions. With dropout, dropout masks m_1 are applied to the encoder layer and m_2 to the decoder layer, as shown in Eqs. 7.4–7.7. Where $\odot$ denotes element-wise multiplication and m_1 and m_2 are binary mask vectors with each element sampled from a Bernoulli distribution; recall that this distribution is characterized by the probability of failure $1 - p$ and success p. During inference (testing), dropout is not applied. Instead, the weights are scaled by $1 - p$ to account for the dropped neurons during training, as seen in Eqs. 7.8 and 7.9.

$$z = f_{\text{encoder}}(W_1 x + b_1) \tag{7.2}$$

$$\hat{x} = f_{\text{decoder}}(W_2 z + b_2) \tag{7.3}$$

$$m_1 \sim \text{Bernoulli}(p) \tag{7.4}$$

$$m_2 \sim \text{Bernoulli}(p) \tag{7.5}$$

$$z = f_{\text{encoder}}((W_1 x + b_1) \odot m_1) \tag{7.6}$$

$$\hat{x} = f_{\text{decoder}}((W_2 z + b_2) \odot m_2) \tag{7.7}$$

$$z_{\text{test}} = f_{\text{encoder}}(W_1 \cdot (1 - p)x + b_1) \tag{7.8}$$

$$\hat{x}_{\text{test}} = f_{\text{decoder}}(W_2 \cdot (1 - p)z + b_2) \tag{7.9}$$

7.1.2 Variational Autoencoders

VAEs are a powerful generative AI model designed to reconstruct input data and generate new data points similar to the training dataset (see Fig. 7.2). VAEs extend the concept of autoencoders by introducing a probabilistic approach to the latent space. Instead of mapping the input data x directly to a single point in the latent space z, in VAE, the encoder maps it to a distribution over the latent space typically represented by a mean μ and variance σ^2, meaning that it maps the input data x to these parameters of the distribution and allows for more controlled and meaningful data generation. The decoder reconstructs the input data by sampling from this distribution and mapping the sample back to the input space, trying to reproduce x. In general, VAEs are particularly useful for tasks that require generating new data points similar to the original dataset, such as image synthesis and text generation.

VAEs can be applied in various functional areas of network management. In fault management, leveraging the VAE's ability to model complex data distributions, they can detect subtle deviations from normal operation that traditional methods might miss. In configuration management, they can help automate the process of network reconfiguration by proposing network configurations that maintain or improve network performance while adhering to constraints and best practices learned from historical configuration data. In accounting management, VAEs can be used to model normal financial transaction patterns within a network. In this regard, when real-time data is fed into the VAE, each transaction that significantly deviates from the expected pattern (as indicated by a high reconstruction error) can be flagged as potential fraudulent activity. In performance management, by sampling the latent space of the VAE, network managers can simulate various traffic scenarios to optimize routing protocols and, thus, improve overall network performance. In security management, similar to anomaly detection, the VAE is trained on normal traffic data, and during operation, it highlights any deviation. A deviation allows for identifying known and unknown network intrusions, making it a powerful tool for enhancing network security.

Algorithm 2 Variational Autoencoder

Input: Dataset D with features X
Output: Generative variational autoencoder model VAE
Procedure:
Initialize the network with random weights and biases
for each epoch **do**
  for each batch of data **do**
    Forward propagate the input through the encoder to compute
    the mean μ and variance σ^2 of the latent space distribution
    Sample from the latent space distribution to obtain a latent vector **z**
    Forward propagate the latent vector **z** through the decoder to reconstruct the input
    Compute the reconstruction loss
    Compute the KL divergence between the learned latent distribution and a prior distribution
    Combine the reconstruction loss and KL divergence to form the total loss
    Backpropagate the error to update the weights and biases using gradient descent
  end for
end for
End Procedure

The VAE (see Algorithm 2) functioning includes:

- Initialize the VAE with random weights and biases.
- Iteratively improve the model by adjusting the weights and biases through forward propagation, computing the loss, and backpropagation.
  - During forward propagation, the input data passes through the encoder to produce the latent space distribution, represented by the mean μ and variance σ^2, as shown in Eqs. 7.10 and 7.11. A latent vector **z** is sampled from this distribution, and the decoder uses **z** to reconstruct the input.

$$\mu = f_{\text{encoder}}(\mathbf{x}) \tag{7.10}$$

$$\sigma^2 = g_{\text{encoder}}(\mathbf{x}) \tag{7.11}$$

- Each neuron in the encoder and decoder behaves as follows. (i) It computes a weighted sum of its inputs. (ii) It adds a bias. (iii) It applies an activation function to introduce nonlinearity into the model, enabling the network to learn complex patterns. (iv) It passes the result of the activation function to the next layer.
- A typical loss function used to guide the optimization process for VAEs is given by Eq. 7.12 that represents a combination of reconstruction loss $\mathbb{E}_{Q(z|x)}[\log P(x|z)]$ and the KL divergence $D_{\text{KL}}(Q(z|x)$. In this equation, $Q(z|x)$ is the distribution of the latent variable z given the input data x, $P(x|z)$ is the probability of reconstructing x given z, and the expectation $\mathbb{E}_{Q(z|x)}$ means averaging all possible values of z sampled from the distribution $Q(z|x)$. Reconstruction loss measures the error between the original and reconstructed input; the VAE agent aims to minimize this error. The KL divergence measures how much the learned latent distribution $Q(z|x)$ deviates from a prior distribution $P(z)$ (a predefined, usually simple, distribution like the standard Gaussian distribution); see Eq. 7.13. If $Q(z|x)$ and $P(z)$ are similar, the KL divergence will be slight; it means that the learned distribution is close to the prior distribution. If $Q(z|x)$ and $P(z)$ are different, the KL divergence will be significant; it indicates that the learned distribution $Q(z|x)$ is significantly different from the prior distribution $P(z)$. In VAEs, the agent aims to learn a distribution $Q(z|x)$ to be close to the prior distribution $P(z)$ because this ensures that the latent space is smooth and organized; a well-behaved latent space allows the VAE to generate meaningful and realistic data.

$$\text{Loss} = \mathbb{E}_{Q(z|x)}[\log P(x|z)] - D_{\text{KL}}(Q(z|x) \parallel P(z)) \tag{7.12}$$

$$D_{\text{KL}}(Q(z|x) \parallel P(z)) = \int Q(z|x) \log\left(\frac{Q(z|x)}{P(z)}\right) \mathrm{d}z \tag{7.13}$$

- During backward propagation, the VAE computes the gradient of the loss function with respect to each weight and bias using the chain rule. Once these gradients are calculated, an optimization algorithm adjusts the weights and biases to minimize the loss function. Typical optimization algorithms for VAEs include SGD, RMSprop, and Adam.

- The final model consists of optimized weights and biases that minimize the total loss function, providing accurate reconstructions and the ability to generate new data points similar to the training dataset.
- VAEs are versatile and can model complex nonlinear relationships in the data by adjusting the encoder and decoder architectures (number of layers and neurons) and hyperparameters.
- VAEs, like other DL models, can be challenging to interpret and require significant computational resources for training, especially with large datasets and deep architectures. They are also sensitive to hyperparameter tuning, requiring careful selection of learning rates, batch sizes, and architectures for optimal performance. Training deep VAEs can lead to issues like vanishing or exploding gradients and overfitting, which can hinder convergence. Normalization, dropout, and advanced initialization techniques often help address these issues.
  - Advanced Xavier Initialization is designed to prevent gradients from becoming either too small (vanishing gradients) or too large (exploding gradients), resulting in more stable and reliable training. Furthermore, proper initialization enables the VAE to start training from a good point, which can lead to faster convergence and improved overall performance.
  - The Xavier Initialization ensures that the variance of the output of each layer is similar to the variance of its input, which helps maintain stable gradients throughout the network. In this regard, in Eq. 7.14, the Xavier Initialization sets the initial weights from a normal distribution $\mathcal{N}$ with mean 0 and variance $\frac{2}{n_{\text{in}}+n_{\text{out}}}$ that depends on the number of input n_{in} and output n_{out} neurons in a layer.

$$W \sim \mathcal{N}\left(0, \frac{2}{n_{\text{in}} + n_{\text{out}}}\right) \tag{7.14}$$

7.1.3 Intrusion Detection

Autoencoder for Intrusion Detection

In security management, intrusion detection is essential for identifying unauthorized access or malicious activity within a network. Autoencoders can detect intrusions by learning the normal patterns of network traffic and flagging deviations from these patterns as potential intrusions. According to the methodology discussed in Chap. 3, for addressing this problem, it is necessary to follow these steps: (i) Collect network traffic data, including features such as packet time delta, source and destination IP addresses, TCP and UDP ports, and packet length. (ii) Generate synthetic data to simulate normal and abnormal network traffic. (iii) Normalize or standardize the features to ensure they are on a similar scale. (iv) Split the dataset into training and testing sets. (v) Train the autoencoder model to learn normal traffic patterns. (vi) Evaluate the model's performance by calculating reconstruction error, which is used to detect intrusions. (vii) Once the model is trained and evaluated, it can be deployed to monitor network traffic and detect intrusions in a real network. List 7.1 shows a Python code snippet to realize the above-described intrusion detection with an autoencoder. This code can be used by readers as a template for dealing with similar anomaly detection problems.

```
#Import Libraries
import numpy as np
import pandas as pd
from sklearn.preprocessing import StandardScaler
from tensorflow.keras.models import Sequential
from tensorflow.keras.layers import Dense
import matplotlib.pyplot as plt
import seaborn as sns

#Generate a synthetic dataset for demonstration
np.random.seed(42)
#Features: "packet.time_delta, ip.src, ip.dst, tcp.srcport, tcp.dstport, udp.srcport, udp.dstport, packet.len"
normal_data = np.random.normal(loc=[0.001, 192168010, 192168011, 80, 8080, 53, 53, 60],
                               scale=[0.0005, 10, 10, 10, 10, 5, 5, 10],
                               size=(1000, 8))
abnormal_data = np.random.normal(loc=[0.1, 192168111, 192168112, 22, 23, 443, 12345, 100],
                                 scale=[0.02, 20, 20, 20, 20, 10, 10, 20],
                                 size=(50, 8))
data = np.concatenate([normal_data, abnormal_data])
np.random.shuffle(data)

#Normalize the data
scaler = StandardScaler()
data = scaler.fit_transform(data)

#Split data into training and testing sets
train_data = data[:900]
test_data = data[900:]

#Build the Autoencoder model
autoencoder = Sequential([
    Dense(8, activation='relu', input_shape=(8,)),
    Dense(4, activation='relu'),
    Dense(2, activation='relu'),
    Dense(4, activation='relu'),
    Dense(8, activation='relu'),
    Dense(8, activation='sigmoid')
])

autoencoder.compile(optimizer='adam', loss='mse')
autoencoder.fit(train_data, train_data, epochs=50, batch_size=16, validation_split=0.2, verbose=1)

#Detect intrusions by calculating reconstruction error (reconstruction loss)
```

```
reconstructions = autoencoder.predict(test_data)
mse = np.mean(np.power(test_data - reconstructions, 2), axis=1)

#Define a threshold for intrusion detection
threshold = np.percentile(mse, 95)
intrusions = mse > threshold

#Plotting the distribution of reconstruction error
sns.histplot(mse, bins=50, kde=True)
plt.axvline(threshold, color='r', linestyle='--', label='Threshold')
plt.title('Reconstruction Error Distribution')
plt.xlabel('Reconstruction Error')
plt.ylabel('Frequency')
plt.legend()
plt.show()

#Print number of detected intrusions
print(f"Number of detected intrusions: {np.sum(intrusions)}")
```

Listing 7.1 Autoencoder - Intrusion Detector

7.2 Adversarial Models

An adversarial model involves two neural networks, a generator and a discriminator, trained simultaneously in a competitive setting. Competitive means that the two networks are trained against each other, continually pushing the generator to improve. The generator aims to minimize the discriminator's ability to detect fake data, while the discriminator seeks to maximize its accuracy in distinguishing between real and fake data. The most well-known examples of this model are the GANs and the adversarial autoencoders.

GANs are a powerful generative AI model designed to create new data samples that closely resemble the training dataset. The architecture of GANs involves two DNNs, namely a generator and a discriminator. These DNNs are trained simultaneously in a competitive setting (see Fig. 7.3). The generator intends to produce data samples that are indistinguishable from real data. The discriminator aims to differentiate between real data and data generated by the generator. Following this adversarial process, the generator improves its ability to create realistic data, and the discriminator enhances its capability to distinguish real data from fake data. GANs have been successfully applied to generate high-quality images from random noise and create deepfake videos. They can also generate additional training data (data augmentation) to improve the performance of ML models in scenarios where labeled data is scarce.

GANs can be applied in several functional areas of network management. In fault management, GANs can generate synthetic fault data to improve the robustness and accuracy of fault detection ML algorithms. In configuration management, GANs can simulate various network configurations, helping network practitioners evaluate the impact of different setups in a controlled environment without disrupting the production network. In accounting management, GANs can generate synthetic data that mimics real-world usage patterns, facilitating more accurate budgeting for investments and better profit forecasting. In performance management, GANs can simulate network traffic under different conditions to enable more comprehensive testing of novel network protocols, routers, and switches. In security management, GANs can generate synthetic attack scenarios that are applicable for evaluating and improving the resilience of network security systems against potential threats. In this area, GANs can also create adversarial examples that test the strength of security algorithms, identifying vulnerabilities before malicious actors can exploit them.

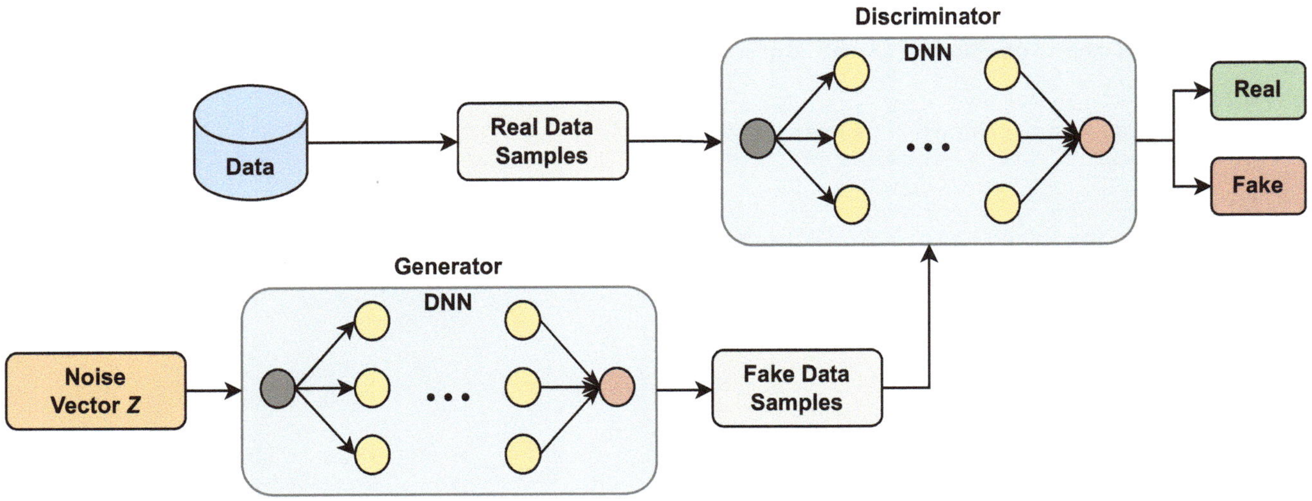

Fig. 7.3 Generative adversarial network architecture

Algorithm 3 Generative Adversarial Network

Input: Dataset D with features X
Output: Trained Generator and Discriminator models
Procedure:
Initialize the Generator and Discriminator networks with random weights and biases
for each epoch **do**
  for each batch of data **do**
    Sample a batch of real data from the dataset D
    Generate a batch of fake data using the Generator
    Forward propagate the real and fake data through the Discriminator to get classification scores
    Compute the Discriminator loss by comparing the real data scores with the fake data scores
    Backpropagate the Discriminator loss and update its weights and biases using gradient descent
    Generate a new batch of fake data using the updated Generator
    Forward propagate the fake data through the Discriminator to get updated classification scores
    Compute the Generator loss based on Discriminator's ability to differentiate real from fake data
    Backpropagate the Generator loss and update its weights and biases using gradient descent
  end for
end for
End Procedure

The GAN (see Algorithm 3) functioning includes:

- Initialize the generator and discriminator with random weights and biases.
- Iteratively improve the models through a two-part process:
  - Discriminator Training. For each mini-batch of training data, the following process occurs: ① The discriminator receives a batch of real data samples x from the training dataset and processes this data to produce a classification score $D(x)$, where $D(x)$ represents the probability that x is an actual sample (as opposed to being generated by the generator). ② The generator generates a batch of fake data samples $\tilde{x} = G(z)$ using random noise vectors z sampled from a prior distribution (usually a uniform distribution). The discriminator then processes this fake data to produce a classification score $D(\tilde{x})$, where $D(\tilde{x})$ represents the probability that $\tilde{x}$ is an actual sample. ③ The discriminator loss $\mathcal{L}_D$ is computed by comparing the discriminator's predictions for both the real and fake data. ④ The discriminator network updates weights and biases by backpropagating the error from the loss $\mathcal{L}_D$; the objective is to minimize $\mathcal{L}_D$, thereby improving the ability of the discriminator to distinguish between real and fake data. Here $\mathcal{L}_D$ is given by Eq. 7.15, where $x \sim p_{\text{data}}$ denotes samples drawn from the real data distribution, $z \sim p_z$ denotes samples drawn from the noise distribution used by the generator, $D(x)$ is the discriminator's output for real data, and $D(G(z))$ represents the discriminator's output for the fake data generated by the generator.

$$\mathcal{L}_D = -\left[\mathbb{E}_{x\sim p_{\text{data}}}[\log D(x)] + \mathbb{E}_{z\sim p_z}[\log(1 - D(G(z)))]\right] \tag{7.15}$$

- Generator Training. After updating the discriminator, the generator improves its ability to generate realistic data that can deceive the discriminator as follows. ① The generator produces a batch of fake data samples $\tilde{x} = G(z)$ using random noise vectors z. ② The discriminator evaluates the fake data and produces a classification score $D(\tilde{x})$. The generator's goal is to maximize this score, meaning that it wants the discriminator to classify the fake data as real. ③ The generator loss $\mathcal{L}_G$ is computed based on the ability of the discriminator to classify the fake data correctly. The generator's objective is to minimize $\mathcal{L}_G$ given by Eq. 7.16. ④ The generator's weights and biases are updated by backpropagating the error from the loss $\mathcal{L}_G$. The objective is to minimize $\mathcal{L}_G$, which effectively improves the generator's ability to generate realistic data that can fool the discriminator.

$$\mathcal{L}_G = -\mathbb{E}_{z\sim p_z}[\log D(G(z))] \tag{7.16}$$

- The GAN training process alternates between discriminator training and generator training in each iteration. Over time, this adversarial process leads to both networks improving: The generator becomes better at producing realistic data, and the discriminator becomes more proficient at distinguishing real data from fake data.

- The final GAN model comprises a trained generator capable of producing realistic data and a discriminator that effectively distinguishes between actual and fake data.
- As mentioned above, GANs are highly versatile and can generate data in various domains, including network management. Nonetheless, training a GAN requires significant computational resources for data generation in complex tasks. Furthermore, GAN training presents challenges related to issues such as mode collapse, where the generator produces a limited variety of data, and training instability, which occurs when the generator and discriminator fail to converge properly. The use of alternative loss functions, careful tuning of learning rates and architectures, and batch normalization can help address the issues in GANs training.

  - Batch normalization aids in addressing the "internal covariate shift" problem: During training, as the inputs to each layer in a DNN change (due to changes in the parameters of the previous layers), the distribution of inputs to each layer can also change; as a result, each layer must constantly adapt to the new distribution, which can slow down the training process and make it difficult for the model to converge. By reducing internal covariate shifts, batch normalization allows the GAN to train faster and with a higher learning rate, often leading to faster convergence. Also, batch normalization makes the network less sensitive to the initial values of the weights, reducing the need for careful initialization.
  - During training, batch normalization operates as follows. The input data is divided into mini-batches. For each mini-batch, the data is normalized by calculating the mean and variance of the inputs for that batch. For each input in the mini-batch, this normalization technique subtracts the mean and divides by the square root of the variance (plus a small constant, called epsilon, to avoid division by zero). This step ensures that the inputs have a mean of zero and a standard deviation of one, effectively normalizing the data, as shown in Eq. 7.17. In this equation, x_i is the input value, μ_{B} is the mean of the inputs in the mini-batch, σ_{B}^2 is the variance of the inputs in the mini-batch, ϵ is a small constant added to the variance for numerical stability, and $\hat{x}_i$ is the normalized input.

$$\hat{x}_i = \frac{x_i - \mu_{\text{B}}}{\sqrt{\sigma_{\text{B}}^2 + \epsilon}} \tag{7.17}$$

  - After normalization, two learnable parameters are introduced, a scale parameter γ and a shift parameter β, which allow the GAN to learn the appropriate scale and shift of the normalized data (see Eq. 7.18). The normalized and scaled inputs y_i are then passed through the subsequent layers of the network. During backpropagation, the parameters γ and β are updated along with the other network parameters to minimize the loss function.

$$y_i = \gamma\hat{x}_i + \beta \tag{7.18}$$

- During inference, batch normalization utilizes the running mean and variance (computed during training) to normalize the inputs, rather than the mean and variance of the mini-batch. Thus, the model's predictions are consistent and not dependent on the particular batch of data.

7.3 Transformers

A transformer is a DNN architecture (model) with self-attention mechanisms; it was introduced by Vaswani et al. in the paper "Attention is All You Need" [1].

7.3.1 Architecture

As mentioned earlier, the transformer architecture (see Fig. 7.4) involves stacks of encoders (and possibly decoders), each consisting of self-attention and feedforward layers. In this vein, a transformer can be encoder-only, decoder-only, or encoder-decoder. In general, the encoder processes the input sequence and the decoder generates the output sequence (used in tasks like language translation).

Pivotal components

- Tokenization is a crucial step in transformers for data understanding, as it allows breaking a sequence of text into smaller parts, called tokens. The tokens are used to build words, and the words are merged to form new terms. The vocabulary of words defines the expression boundary of a transformer. Different tokenizers exist, varying in the level of granularity of the text breakdown, from breaking it down into single words to breaking it down into characters or even smaller units. Typical tokenizers are BytePairEncoding, WordPieceEncoding, and Unigram tokenization.
  - BytePairEncoding is a subword tokenization technique conceived for data compression and adapted for NLP tasks. This technique operates as follows. ① It starts with a vocabulary of all individual characters in the text. ② It identifies all pairs of adjacent tokens (initially characters) in the text and counts their frequencies. ③ It finds the most frequent pair of tokens and merges them into a new token; each found token is added to the vocabulary. ④ It repeats steps ② and ③ until to create a fixed-size vocabulary of subword units. This technique efficiently handles rare and out-of-vocabulary words by breaking them into smaller, known subwords. It is also adaptive to different languages by learning the most frequent subwords in a given corpus. Once the vocabulary is set, it needs to be retrained to adapt to new data.
  - WordPieceEncoding is a subword tokenization technique focused on maximizing the likelihood of the tokenized sequence under a given training corpus. This technique operates as follows. ① It starts with a base vocabulary of individual characters and special tokens. ② It adds subword units to the vocabulary by choosing the one that maximizes the likelihood of the tokenized sequence in training data, given the current vocabulary. ③ For a given word, it selects the longest matching subword in the vocabulary. ④ It splits the word into subwords based on the vocabulary, usually starting with the longest possible subwords. As BytePairEncoding, WordPieceEncoding handles rare words by breaking them down into common subwords. Moreover, WordPieceEncoding achieves context-awareness by selecting subwords that maximize the likelihood of the text. This technique can be complex computationally when determining which subwords maximize the probability of tokenized data.
  - Unigram is a probabilistic subword tokenization technique that operates as follows. ① It starts with an extensive vocabulary that includes all possible subwords (words and individual characters). ② It assigns each subword a probability of appearance. ③ It prunes subwords with lower probabilities to reduce the vocabulary size and retain the most relevant subwords. ④ For each word, Unigram selects the most probable subword sequence from the pruned vocabulary. On the one hand, using probabilities enables Unigram to perform more nuanced tokenization, which

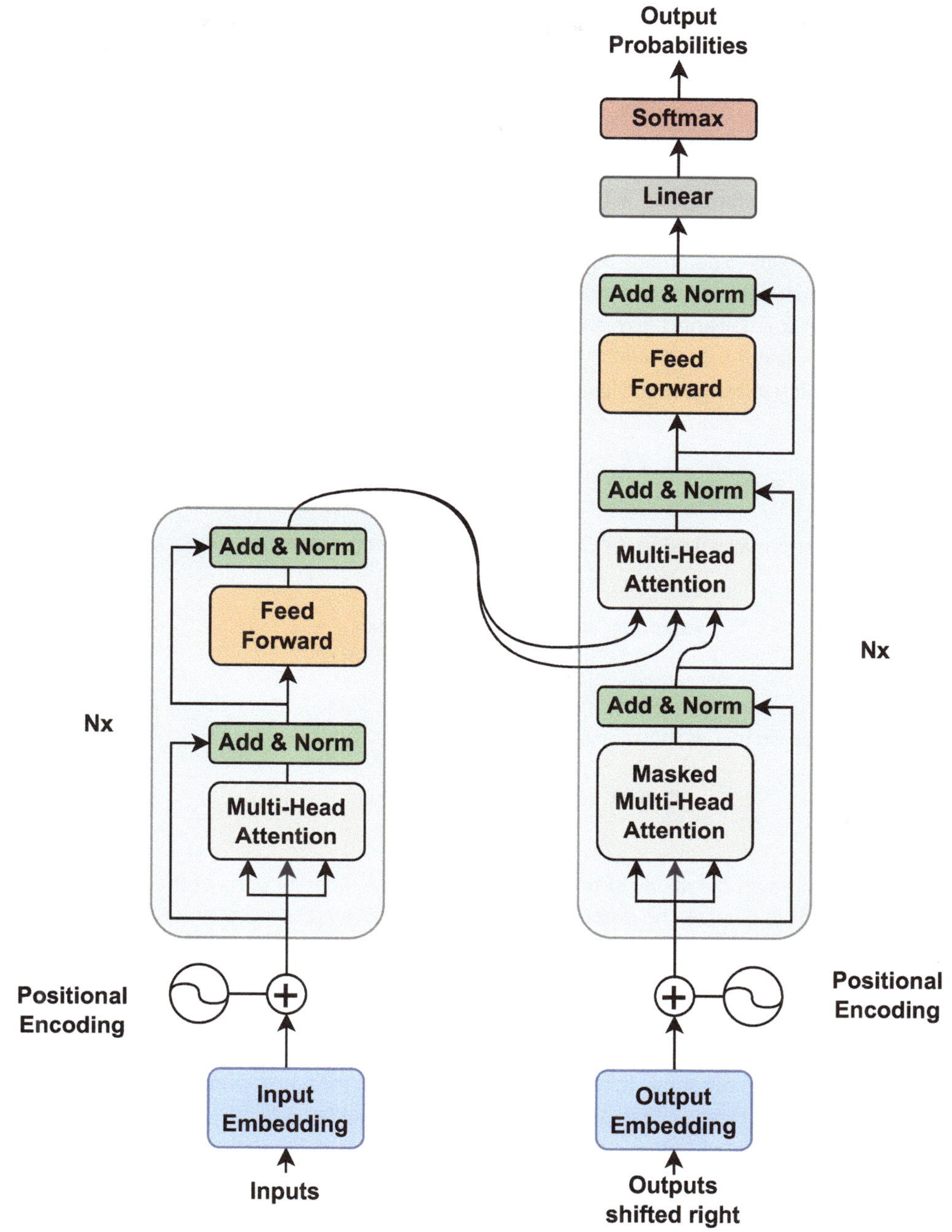

Fig. 7.4 Transformers architecture

can capture variations in language. On the other hand, as with WordPieceEncoding, the probability computation leads to the Unigram technique having a high computational complexity.

- The self-attention mechanism is the core of the transformer architecture. This mechanism transforms a sequence of tokens X, as shown in Eq. 7.19, where x_i is an individual token (i.e., the unit of text that a model process, such as word, subword, or character) and n is the total quantity of tokens, into a new representation formed by queries, keys, and values. Equation 7.20 exemplifies a transformation, where $Query(Q)$ represents what the token is searching for in the sequence, $Key(K)$ represents the content of the token, $Value(V)$ holds the actual data to be used in the attention computation, W_Q, W_K, W_V are learnable weight matrices, typically $\in \mathbb{R}^{d \times d_k}$ - d is the embedding dimensionality.

$$X = [x_1, x_2, \ldots, x_n] \tag{7.19}$$

$$
\begin{aligned}
Q &= XW_{\mathrm{Q}} \\
K &= XW_{\mathrm{K}} \\
V &= XW_{\mathrm{V}}
\end{aligned}
\tag{7.20}
$$

- The attention score is pivotal in the self-attention mechanism. This score determines how much focus (or "attention") a token in a sequence should give to other tokens, based on the similarity of their query and key vectors. Equation 7.21 computes the attention score for token i with respect to token j in a sequence. In this equation, Q_i represents the query vector corresponding to the i-th token in the sequence, K_j is the key vector corresponding to the j-th token in the sequence (this vector aids to determine how relevant the token is to the queries from other tokens), and $Q_i \cdot K_j^T$ represents the dot product between the query vector Q_i for token i and the transpose of the key vector K_j for token j (this product measures the similarity between the two vectors; a higher value indicates that token j is more relevant to token i). Furthermore, d_k is the dimension of the key vector, and $\frac{1}{\sqrt{d_k}}$ helps stabilize the gradient by preventing the dot products from growing too large which leads to better learning during training.

$$
\text{Attention score}(i, j) = \frac{Q_i \cdot K_j^T}{\sqrt{d_k}} \tag{7.21}
$$

- The attention scores are normalized using the Softmax function to get the attention weights, as shown in Eq. 7.22.

$$
\text{Attention\_weights}(i, j) = \text{softmax}\left(\frac{Q_i \cdot K_j^T}{\sqrt{d_k}}\right) \tag{7.22}
$$

- The attention weights are used to compute a weighted sum of the value vectors. The output of each token is the weighted sum of all value vectors, as shown in Eq. 7.23.

$$
\text{Output}_i = \sum_j \text{attention\_weights}(i, j) \cdot V_j \tag{7.23}
$$

- Multi-head attention improves the ability of transformers to focus on different parts of the input simultaneously by applying attention heads in parallel. The output of each head is concatenated according to Eq. 7.24, where W_O is a learnable weight matrix.

$$
\text{MultiHead}(Q, K, V) = \text{Concat}(\text{head}_1, \ldots, \text{head}_h)W_{\mathrm{O}} \tag{7.24}
$$

- After self-attention, each position is independently passed through a layer of FNNs to help the model to learn richer representations of tokens and their context, see Eq. 7.25. In this equation, x is the output from the self-attention mechanism for a given token (i.e., a single token's embedding or an intermediate representation after passing through earlier layers), W_1 is a weight matrix that defines a linear transformation applied to the input x (i.e., a learnable matrix during training with dimension $d_{\text{model}} \times d_{\text{hidden}}$, where d_{model} is the dimension of the input and d_{hidden} correspond to the dimension of the hidden layer), b_1 is a bias term that has the same dimension that the hidden layer), $\max(0, xW_1 + b_1)$ applies ReLU to the intermediate result to introduce nonlinearity into the model and, thus, allow the network to learn more complex patterns, W_2 is another weight matrix with dimension $d_{\text{hidden}} \times d_{\text{model}}$ that transforms the output of the ReLU activation into a final output vector that projects the hidden layer back to the same dimensionality as the input x), and b_2 is another bias term.

$$
\text{FFN}(x) = \max(0, xW_1 + b_1)W_2 + b_2 \tag{7.25}
$$

- The transformers apply the concepts of residual connection and normalization (see Eq. 7.26) to the output of their architectural sublayers (e.g., self-attention mechanism or FNN). The residual connection ($x + \text{Sublayer}(x)$) prevents the model from deviating too far from the original input x; it adds the transformation performed by the sublayer to the original input to avoid the model from losing important information. Normalization ensures that the combined output is stable and normalized across the transformer, which helps stabilize and accelerate the training process, allowing the transformer to learn more effectively.

$$
\text{Output} = \text{LayerNorm}(x + \text{Sublayer}(x)) \tag{7.26}
$$

- Transformers use positional encoding to add order information to the input embeddings; see Eq. 7.27 where pos represents the position of the token in the sequence, i refers to the dimension of the encoding vector, d is the total dimension of the encoding vector (usually the same as the dimension of the embedding or input vectors to the model), and $10{,}000^{2i/d}$ scales the input to ensure that each dimension in the positional encoding vector varies at different frequencies (lower dimensions varying slowly and higher dimensions varying quickly). The sine and cosine functions introduce periodic variations (interpreting the position as an angle in a high-dimensional space), which allows the transformer to learn relationships between positions.

$$PE_{(\text{pos},2i)} = \sin\left(\frac{\text{pos}}{10{,}000^{2i/d}}\right)$$
$$PE_{(\text{pos},2i+1)} = \cos\left(\frac{\text{pos}}{10{,}000^{2i/d}}\right) \quad (7.27)$$

- Pre-training is the first learning step typically performed by a transformer to acquire the fundamentals of language understanding; it utilizes language modeling techniques (e.g., autoregressive and masked) that enable an in-depth and versatile understanding of the language. Posteriorly, training allows transformers to comprehend and generate human-like text with coherence by learning the probability distribution of the vocabulary from a vast corpus using diverse techniques, such as next-token prediction, masked language modeling, n-grams, and continuous space modeling.

Large Language Model and Short Language Model

In general, modern Small Language Models (SLMs) and Large Language Models (LLMs) are based on the transformer architecture, which, in turn, is based on the self-attention mechanism to perform NLP tasks efficiently. Their key differences are related to goal, size, and privacy.

- LLMs provide generalized support for a wide range of NLP tasks (e.g., text completion, translation, summarization, and question answering) across diverse application domains. Unlike the generalization sought by LLMs, SLMs aim to assist in the realization of more specific or straightforward NLP-assisted tasks (e.g., detecting intrusions in a tokenized dataset containing samples of normal and abnormal traffic) in a particular application domain, such as network management.
- About the size, it is essential to differentiate the scale of parameters and layers used by models. In this regard, an LLM considers millions to billions of parameters (or more), while an SLM operates with thousands or millions of parameters. Leveraging the size reduction, an SLM offers responses faster and demands fewer computational resources than an LLM. SLMs can be used in more constrained environments than LLMs. Conversely, thanks to its size, an LLM can capture complex patterns and nuances in a natural language corpus that an SLM cannot.
- Achieving data privacy is more challenging in an LLM than in an SLM, as handling sensitive information in a cloud-deployed LLM requires strict access control and encryption. Processing sensitive data in a centralized setting can expose organizations to privacy issues when the model itself or the infrastructure is compromised. In SLMs, which are usually deployed locally, private data remains within the organization, reducing exposure to external threats. Remarkably, despite an SLM being more privacy-aware than an LLM, an SLM may fail to identify complex, multi-layered threats that require deep contextual understanding, which LLMs might handle more effectively.

7.3.2 Encoder-Only

Encoder-only transformers operate exclusively with encoders and use the self-attention mechanism to capture dependencies between different words, regardless of their position, by considering only the input data. An encoder-only transformer comprises a stack of layers. ① The input embedding layer. ② The positional encoding layer. ③ The self-attention layers (bidirectional). ④ The FNN layers. ⑤ The layer of residual connections and normalization. ⑥ The output layer (optional depending on the task) is, in many cases, passed to a task-specific output layer to, for example, predict labels or produce class probabilities for each token generated by the encoder. About these architectural layers, it is remarkable that the self-attention mechanism in encoder-only is bidirectional, meaning that each token in the input sequence attends to every other token, both before and after. Furthermore, since the model requires seeing the entire input sequence for deep understanding, there is no need to mask tokens from future positions.

BERT [2], Robustly optimized BERT approach (RoBERTa) [3], DistilBERT [4], ALBERT [5], and Gemini [6] exemplify LLMs and SLMs following the encoder-only architecture. This architecture generally excels at learning bidirectional context, making it well-suited for tasks that require a deep understanding of the entire input sequence, rather than just generating the next token, as in decoder-only architectures. From a broad perspective, the encoder-only models are appropriate for performing text classification tasks.

Large and Small Models

- BERT [2] is a well-known encoder-only LLM devised in 2018 by Google for language understanding tasks (e.g., question answering and text classification). This encoder-only LLM, including its versions BASE (110 M parameters) and LARGE (340M parameters), is characterized by two pre-training tasks: Masked Language Modeling (MLM) and Next Sentence Prediction (NSP) techniques. MLM supports BERT in getting a deep understanding of context and relationships between words by learning to fill in missing information and improving language comprehension and representation abilities by enabling the prediction of masked tokens based on the contextual information provided by the other unmasked tokens; a masked token is a token replaced with a MASK. For instance, in the sentence "a flow with a size larger than 10K is a MASK," the model must learn to predict that the masked token should be "elephant" based on the context that the surrounding words offer. With NSP conceived to assist a model to learn the coherence and logical flow between pairs of sentences, BERT learns sentence-level relationships; it can predict whether the second sentence follows the first one in a sequence of tokens. For instance, if the sentence "a flow with a size larger than 10K is a MASK" is followed by the sentence "the router is suffering a service outage," the last sentence is labeled as NotNext.
- RoBERTa [3] is a BERT variant introduced by Facebook in 2019, which has been trained on much larger datasets for more extended periods and with more computational resources, leading to improvements in performance on various language understanding tasks. RoBERTa is an LLM with 335 M parameters that preserves the encoder-only architecture of BERT, but in the pre-training process, it does not use NSP (experiments revealed that it was not necessary for performance gains) and performs dynamic masking (masking pattern changes during training) instead of static masking.
- DistilBERT [4] is an SLM introduced by Hugging Face in 2019. This SLM is smaller (40%) and faster (60%—66 M compared to 110 M of parameters) than BERT; notably, DistilBERT maintains BERT's learning capability (97%) through the knowledge distillation technique. In general, this technique allows for transferring knowledge from a teacher model (more extensive and complex) to a student model (a smaller and simpler model) to retain the performance and capabilities of the larger model while reducing computational complexity, making the smaller model faster and more efficient for practical use. Notably, during the knowledge distillation process, the student model is trained on the original labeled data and the soft targets produced by the teacher model. Soft targets are output probabilities that contain more information than hard targets (the correct labels in a classification task); for instance, the student can learn from the teacher that a flow has an 80% probability of being an elephant and a 20% likelihood of being a mouse.
- ALBERT [5], proposed by Google Research in 2019, is an SLM-based BERT with 60M parameters. ALBERT utilizes Factorized Embedding Parameterization, Parameter Sharing, and Sentence Order Prediction to achieve a smaller and faster model than BERT for inference tasks. With Factorized Embedding Parameterization, ALBERT splits the large vocabulary embedding matrix into two smaller matrices to decrease the number of parameters in the embedding layer. By Parameter Sharing, ALBERT can share parameters across layers in the transformer stack, reducing the number of parameters. ALBERT uses Sentence Order Prediction instead of NSP to accurately predict the order of two consecutive sentences, thereby capturing intersentence coherence more effectively.

- Gemini [6], introduced by Google, is a multi-modal encoder-only LLM enriched with machine reasoning. This LLM is helping to address tasks such as text and code generation, reasoning, and image interpretation. For the sake of the trade-off between computational resource demands and performance, there are various Gemini flavors: Gemini Nano (with 1.8 B and 3.25 B parameters), Gemini Pro (size non-officially disclosed up to now), and Gemini Ultra, which is the largest model, but its size has also not been officially disclosed.

7.3.3 Decoder-Only

Decoder-only transformers operate exclusively with decoders and generate coherent text by predicting the following words in a sequence, considering previously generated words using the self-attention mechanisms (e.g., summarization and translation tasks). A decoder-only transformer comprises a stack of layers. ① The input embedding layer. ② The positional encoding layer. ③ The masked self-attention layers (unidirectional). ④ The FNN layers. ⑤ The layer of residual connections and normalization. ⑥ The output linear layer projects tokens into the vocabulary space; the projections are transformed into probabilities representing how much each token in the vocabulary can be the next token in the sequence. Regarding these architectural layers, it is remarkable that the masking and unidirectional nature of the attention mechanism (tokens only attend to past tokens) make the decoder-only architecture autoregressive; it can generate tokens one at a time based on the previous tokens.

GPT-series [7], DialogGPT [8], GPT-2 Small [9], GPT-Neo [10], and GPT-J [11] exemplify LLMs and SLMs following the decoder-only architecture. From a broad perspective, these models are helping in facing generative text tasks. Furthermore, various of them are initially trained on an extensive collection of diverse, unlabeled texts in an unsupervised learning mode, followed by discriminative fine-tuning for each specific task.

Large and Small Models

- The GPT series (GPT-2, GPT-3, GPT-4, and GPT-5), introduced by Open AI, are decoder-only LLMs designed for language modeling and a wide range of text generation tasks. GPT-5 and GPT-4 [12] are large-scale, autoregressive LLMs with more than 1T parameters that generate human-like text based on input prompts and represent state-of-the-art advancements in NLP. These LLMs are characterized by their capability to handle multi-modal inputs (they can handle tasks with text and image inputs). Furthermore, thanks to their size, GPT-5 or 4 provides more accurate and coherent responses than their predecessors when dealing with complex tasks.
- GPT-Neo [10] is an open-source GPT-3-based LLM proposed by EleutherAI to democratize LLM access. In this vein, GPT-neo is available for the community to use and modify, making it a valuable resource for developers and researchers. This LLM comes in two flavors that support handling generative tasks: GPT-Neo with 1.3 billion parameters and GPT-Neo with 2.7 billion parameters. Note that GPT-Neo is smaller than GPT-3, which has 175 billion parameters, but still performs well. GPT-J [11] is another open-source LLM developed by EleutherAI, aiming to balance performance and computational efficiency when coping with text generation tasks. This LLM is smaller than GPT-3 but bigger than GPT-Neo; GPT-J considers 6 B parameters.
- GPT-2 Small [9] is an SLM version of GPT-2. This SLM operates with 117M, while GPT-2 functions with 1.5 B parameters. Thanks to its reduced size, GPT-2 Small decreases computational requirements but struggles with complex tasks and the coherence of its responses.
- DialoGPT [8] is a GPT-2-based SLM introduced by Microsoft. This SLM is specifically fine-tuned on high-quality dialogue datasets (trained on 147 million conversation-like exchanges extracted from the Reddit social media network). Therefore, it is suitable for chatbot applications, as it enables the generation of human-like conversa-

tional responses. DialoGPT-small with 117 million, DialoGPT-medium with 345 million, and DialoGPT-large with 762 million parameters are DialoGPT flavors conceived to offer a trade-off between computational resources and performance.

7.3.4 Encoder-Decoder

Encoder-decoder transformers, such as T5 [13] and BART [14], combine the encoder to process the input and the decoder to generate the output. This type of transformer allows performing sequence generation tasks and functions as a unified model with natural language generation and understanding capabilities. The models following this architecture are very efficient in reconstructing sentence fragments; the encoder takes an input sentence with a randomly masked fragment, and the decoder predicts the masked fragment.

Large and Small Models

- T5 [13] is a decoder-encoder and sequence-to-sequence model introduced by Google for addressing NLP tasks (e.g., translation, summarization, classification, and question-answering) in a text-to-text way. This LLM has various flavors seeking to offer a trade-off between computational demands and performance: T5-Small with 60 million parameters, T5-Base with 220 million parameters, T5-Large with 770 million parameters, T5 with 3 billion parameters, and T5 with 11 billion parameters. Remarkably. T5-Small is considered an SLM. In turn, the other flavors are LLMs.
- BART [14] is an LLM, introduced by Facebook, that combines a bidirectional encoder (it can attend tokens in forward and backward directions, as BERT does) and an autoregressive decoder (as GPT does). In this vein, it is suitable for tasks that demand an understanding of the complete input processed by the bidirectional encoder and the generation of coherent outputs offered by the autoregressive decoder. BART includes various sizes by seeking a trade-off between computational resource demands and performance. The most well-known are BART-Base, which operates with 140 million parameters, and BART-Large, which operates with 400 million parameters.

7.3.5 Fine-Tuning and Prompt Engineering

The transformer-based models (i.e., models that follow the architectures discussed earlier: encoder-only, decoder-only, or encoder-decoder) help handle tasks related to the network management functional areas. Beyond the data preprocessing techniques, the performance of these models during task execution can be impacted by the expertise of network practitioners acting as prompt engineers. In this regard, fine-tuning and prompt techniques are necessary to ensure that model customization is accurate, effective, and timely in performing specialized tasks.

Fine-Tuning

Tuning a transformer-based model finely demands turning a general model into a specialized one by training it on specific and smaller datasets to refine its capabilities and performance on particular tasks. The capacity to tune transformers encourages their use in complex and diverse domains, including networking. In the network management domain, fine-tuning can be especially useful for learning the characteristics of flow traffic in a data center, enabling more precise and effective adaptation to the network's specific requirements. Transfer learning,

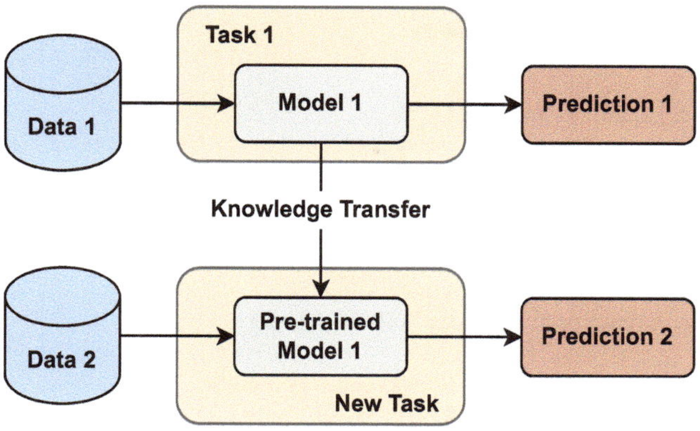

Fig. 7.5 Transfer learning

instruction-tuning, and alignment-tuning exemplify fine-tuning techniques, which can be used individually or in combination to improve the performance of transformer-based models.

- In the Transfer Learning technique (see Fig. 7.5), a general or specific model (e.g., a pre-trained model, such as BERT and GPT-*) is used to build up another one for a different but related task. In this regard, this technique, from a broad perspective, operates as follows. Initially, the pre-trained model is uploaded. After, the parameters of the pre-trained layers are frozen (optionally) to avoid their updating during training. Then, the final fully connected layer is replaced to perform a specific task (e.g., classifying traffic flows). Finally, the model is fine-tuned in the particular dataset by training the new layer. List 7.2 shows a Python code snippet to realize the above-described process using DistilBERT as a pre-trained model to classify normal and abnormal traffic. Readers can use this code as a template for realizing similar transfer learning tasks.

```
#Import libraries
import torch
import torch.nn as nn
from transformers\index{Transformer} import DistilBertForSequenceClassification, DistilBertTokenizer
from torch.optim import Adam
from torch.utils.data import DataLoader

#Load a pre-trained DistilBERT model
#Replace the final classification layer with the from_pretrained method)
#Specify the number of output classes using the num_labels=2 argument - normal and abnormal.
model = DistilBertForSequenceClassification.from_pretrained('distilbert-base-uncased', num_labels=2)

#Freeze the parameters of the pre-trained model
for param in model.distilbert.parameters():
    param.requires_grad = False

#Define loss function and optimizer
criterion = nn.CrossEntropyLoss()
optimizer = Adam(model.parameters(), lr=0.001)

#Assume train_loader is predefined and contains tokenized inputs and labels
#The train_loader in this case represents the tokenized smaller dataset.
for epoch in range(num_epochs):
    model.train()
    for batch in train_loader:
```

```
            # Get input_ids and attention_mask from batch, along with labels
            input_ids = batch['input_ids']
            attention_mask = batch['attention_mask']
            labels = batch['labels']

            #Forward pass
            outputs = model(input_ids=input_ids, attention_mask=attention_mask, labels=labels)
            loss = outputs.loss\xm{\para}

            #Backward pass and optimization
            optimizer.zero_grad()
            loss.backward()
            optimizer.step()
```

Listing 7.2 Transfer Learning

- The instruction-tuning technique refers to fine-tuning a pre-trained model with a dataset of instructions (prompts) and corresponding outputs (answers). Therefore, the model learns to follow user instructions more accurately through the instruction-output pairs. About this technique, it is important to note. First, its fundamental element is the new instructional dataset. Second, it is promising, for example, to assist network practitioners in carrying out simple network troubleshooting via conversational transformer-based models.
- The alignment-tuning process involves adjusting a pre-trained model to adhere to user guidelines related to, for instance, human and ethical values, and operational principles, aiming to ensure that the model behaves according to end-user expectations. Similar to instruction-tuning, this technique first requires a sample dataset; however, in this case, it is used to prevent generating responses with harmful or biased language. Second, it is also a promising technique for helping network practitioners (end-users) via conversational transformer-based models.

Prompt-Engineering

A prompt is a statement or question that guides the output of a transformer-based model. The prompt is helpful to set the context and specify the task(s) that the model must perform. A prompt can play various roles: system, assistant, and user.

- Prompt acting in the system role aims to ensure that the model accurately understands the network practitioner's request and generates a relevant and coherent response. An example of a system role prompt is: "You are an expert in troubleshooting network issues. Your task is to diagnose the root cause of the network problem based on the information provided by the user and suggest a solution to the problem."
- In the assistant role, a prompt specifies a desired task based on the network practitioner's input, understands the context of the practitioner's request, and tailors the response accordingly. The following sentence exemplifies an assistant role prompt: "The root cause of the network issue appears to be problem. To resolve this, you should solution. If the problem persists, consider checking the physical connections and ensuring that the network cables are securely connected."
- A prompt in the user's role (for instance, a non-expert network administrator) encompasses individual interactions with the model, where the user inputs queries, instructions, or commands to receive responses or perform tasks. An example of a user role prompt is: "My internet connection is very slow, and some websites are not loading at all. I've tried restarting my router, but the issue persists. What could be causing this problem?"

Prompt engineering refers to creating prompts intended to improve the quality (in terms of metrics such as accuracy) and applicability (in terms of usefulness and timely delivery) of responses offered by models following transformer architectures. Priming, zero-shot, one-shot, few-shot, chain-of-thought, and self-consistency are prompt engineering techniques.

- Priming involves sending the context to the model before performing a task. For instance, "You are a language model trained to provide detailed and accurate answers. Please always begin your responses with a summary before providing a thorough explanation."
- In zero-shot, no example is passed to the model for the task to perform, as follows: "Translate the following network policy to an Intent Language: keep all links of the network with a delay lower than 10ms."
- In one-shot, a single example is used to show the model what is expected to do. "Please, the command to get the IP Address in linux is ifconfig. Now, tell me what is the command to get the IP Address in windows."
- In few-shot, a list of examples (3 to 5) are sent to show the model what is expected to do. Example1: Translate "Get IP Address" into a linux command: "ifconfig." Example 2: Translate "Check connectivity between two nodes" into a linux command: "ping." Example 3: Translate "Obtain the network name of a host" into a linux command: "hostname." Now, translate the following sentence into a linux command: "Traces the path to a specified destination." Answer: ***Here the model must provide its response according the 3 examples. i.e, traceroute.***
- Chain-of-thought allows the model to reason through the problem to solve, step by step, leading to accurate answers. Let's compute the packet loss step by step. First, obtain with a ping command the $pakects\_sent$ and $packets\_received$. Next, subtract $pakects\_received$ from $pakects\_sent$. Finally, divide the outcome by $pakects\_sent$.
- Self-consistency, it is to generate multiple solutions to a problem and select the most consistent answer based on reasoning or patterns.

  - Prompt-1. Query: A network switch supports 48 ports. If it is upgraded to support 25% more ports, how many ports does it support after the upgrade? Answer: Initially, the switch supports 48 ports. A 25% increase means it supports an additional $48 \times 0.25 = 12$ ports. The total number of ports after the upgrade is $48 + 12 = 60$ ports. The answer is 60.
  - Prompt-2. Query: A data center has 80 servers. If 10% of these servers are down for maintenance, how many servers are operational? Answer: The data center has 80 servers. If 10% are down, then $80 \times 0.10 = 8$ servers are down. Therefore, the number of operational servers is $80 - 8 = 72$ servers. The answer is 72.
  - Prompt-3. Query: A network administrator needs to allocate 2000 GB of bandwidth equally across 8 different institutions. How much bandwidth does each institution receive? Answer: The total bandwidth is 2000 GB. Dividing this equally among 8 institutions means each institution receives $2000 \div 8 = 250$ GB. The answer is 250 GB.
  - Prompt-4. Query: A company's router can handle 1500 simultaneous connections. After a firmware update, its capacity increases by 20%. How many simultaneous connections can the router handle now? Answer: The router initially handles 1500 connections. A 20% increase means the capacity is now $1500 \times 0.20 = 300$ more connections. The total capacity is $1500 + 300 = 1800$ connections. The answer is 1800.
  - Prompt-5. Query: A network uses 120 IP addresses for its devices. If 40% of these addresses are reserved for future use, how many IP addresses are available for current devices? Answer: The network has 120 IP addresses. If 40% are reserved, then $120 \times 0.40 = 48$ IP addresses are reserved. Therefore, the number of IP addresses available for current devices is $120 - 48 = 72$. The answer is 72.
  - Prompt-6. Query: A network practitioner has 80 routers to configure. If 25% of these routers must be reserved for future use, how many devices has the practitioner available for current usage? Answer: ***Here the model must provide its response according the reasoning teaching by the 5 prompts previous.***

7.3.6 Large and Small Models in Network Management

SLMs and LLMs, following the encoder-only, decoder-only, and decoder-decoder architectures, can be utilized to carry out network management tasks framed within the FACAPS. In general, regarding the use of SLMs and LLMs in the functional areas of network management, it is remarkable that, first, SLMs can miss nuanced or complex patterns (such as attacks, configurations, and failures) that LLMs can identify. Second, as LLMs require substantial computational resources, they are challenging to deploy in resource-constrained environments, such as the IoT-Edge continuum, where SLMs could be more suitable. Third, LLMs are more susceptible to adversarial attacks than SLMs. Fourth, it is necessary to answer fundamental questions: are network management practitioners using a cannon to kill a fly? How can they address the hallucination problem brought on by SLMs and LLMs? Can they explain the outcomes produced by large and small models?

Faults Management

This functional area of network management can benefit from the capabilities of LLMs and SLMs to analyze large, multi-modal, and multi-dimensional data. In this regard, LLMs based on the encoder-only architecture can, for instance, be effectively leveraged to identify the root cause, localize, and diagnose failures in extensive network data logs generated by small, medium, and large networks. Beyond reactive fault management, applications that proactively handle faults can benefit from the generative capacity of LLMs, particularly when utilizing decoder-only or encoder-decoder architectures. Data augmentation for improving predictive fault analysis is fundamental to building modern network self-healing management systems. On the other hand, by tokenizing traditional network-fault datasets and fine-tuning LLMs, SLMs based on the encoder-only architecture can be generated to carry out binary and multi-classification tasks. In this vein, potential applications of SLMs include traffic classification in both normal and anomalous conditions, as well as according to the type of failure. In addition, conversational-centered SLMs are pivotal for offering effective basic help desk resolution and service desk delivery, which can decrease operational expenditures and improve QoE.

Configuration Management

This functional area of network management can leverage the capabilities of LLMs and even SLMs to analyze and generate coherent, accurate data. In this sense, models (SLM and LLM) based on the encoder-only architecture can be used as pivots to build up sophisticated configuration systems with the capacity to check and verify the syntax and security of network settings. In addition, similar to coding practices improved by generative AI, data augmentation capabilities offered by SLMs and LLMs, following the encoder-decoder architecture, can help enhance self-configuration by facilitating the creation of autonomous generators and verifiers of human-intelligible and actionable network configurations, considering the network status, external templates, and device knowledge. On the other hand, after tokenizing datasets containing samples of network configurations and fine-tuning an LLM, IDN can benefit from the capability of SLMs based on decoder-only and decoder-encoder architectures to perform specific generation tasks to translate network goals expressed in natural language to business or resource intents to facilitate the turning of traditional networks into self-driving networks, which deliver on business and sustainability targets. Here, SLMs can also be trained to translate intents from business to service, service to resource, and resource to actionable device commands, and vice versa.

Accounting Management

This functional area of network management can benefit from the ability of LLMs and SLMs to analyze large, multi-modal, and multi-dimensional data. LLMs can be used to create and evaluate new business and pricing models by considering internal and external data. Those models are pivotal in estimating returns on investment and forecasting expected profits. An SLM based on the encoder-only architecture can be built to detect fraud from specific datasets containing samples of users carrying out regular and abnormal consumption of a particular service.

Performance Management

This functional area of network management can exploit the capabilities of LLMs to analyze and generate coherent and accurate data. Encoder-based LLMs can help segment users. Segmentation based on bandwidth, access control, and resource availability can help enhance the network's performance, security, and support for multiple tenants. Additionally, LLMs operating with encoder-only capabilities can integrate monitoring features that accurately and effectively analyze a vast amount of data to detect traffic data patterns that degrade network performance. Moreover, LLMs following the decoder-only and encoder-decoder architectures can help create a self-optimization system that receives feedback from past interventions and considers future optimizations to understand and tune the network performance. On the other hand, an SLM can be trained in the particular dataset of a critical network component (device or link) to identify patterns that typically precede its performance degradation or total outage. SLMs can also classify traffic based on various criteria and adjust the network settings to ensure that critical services receive sufficient bandwidth, thereby enhancing overall network performance.

Security Management

This functional area of network management can benefit from the capacity of LLMs and SLMs to analyze large, multi-modal, and multi-dimensional data. In this regard, LLMs based on the encoder-only architecture can, for instance, be effectively leveraged to find, localize, and diagnose security attacks in extensive network data logs from small, medium, and large networks. Beyond reactive security management, applications that proactively avoid security attacks can benefit from the generative capacity of LLMs, particularly when utilizing decoder-only or encoder-decoder architectures. Data augmentation for improving the handling of zero-day attacks is fundamental to accomplishing modern management systems with network self-healing and self-protection capabilities. On the other hand, by tokenizing traditional network-fault datasets and fine-tuning LLMs, SLMs based on the encoder-only architecture can be generated to carry out binary and multi-classification tasks. In this vein, potential uses of SLMs include traffic classification in normal and abnormal situations, as well as threat detection based on the type of attack (e.g., DoS, DDoS, and ransomware). Additionally, conversational-centered SLMs are essential for providing practical support in the event of cyberattacks.

7.4 Summary

This chapter presented generative models from the network management domain. Initially, it discussed and exemplified the use of autoencoders and Variational autoencoders to perform tasks framed into the FCAPS. Then, it presented GANs and illustrated their use for managing networks. Finally, it introduced the transformers' architectures (encoder-only, decoder-only, and encoder-decoder) and showed how they can be employed in the functional areas of network management.

Homework Problems and Questions

7.1 Explain how the autoencoders and variational autoencoders operate. Also, highlight similarities and differences.

7.2 Create a conceptual map to remark transformers' foundations.

7.3 Classify flows with AE, VAE, and RNN in the UNIV1 dataset available in the book's repository. Compare the models regarding accuracy, balanced accuracy, precision, recall, and F1-score.

7.4 Classify flows in the UNIV1 dataset available in the book's repository by using BERT and DistilBERT. Compare the models regarding accuracy, balanced accuracy, precision, recall, and F1-score.

7.5 Write FCAPS-centered intents in NLP and translate to policies in the format (If-Then-Do) using prompt engineering.

Solutions of This Chapter

7.1 Problem. Explain how the autoencoders and variational autoencoders operate. Also, highlight similarities and differences.

Solution. Autoencoders follow a particular neural network's valuable architecture for tasks like unsupervised learning, dimensionality reduction, denoising, and feature extraction. They aim at learning a compressed representation of the input data and are composed of an encoder responsible for mapping the input data x to a latent space z and a decoder in charge of reconstructing the input data $\hat{x}$ from the latent representation z. The encoder and decoder are trained simultaneously to minimize the reconstruction error, typically using a loss function such as MSE. Variational autoencoders are an autoencoders extension that is valuable for tasks like generative modeling, data synthesis, and unsupervised learning. They aim to learn a probability distribution of the latent space z rather than a deterministic representation. In addition to the reconstruction error, variational autoencoders consider KL divergence to ensure the learned latent distribution $q(z|x)$ is close to a prior distribution $p(z)$ (usually a standard normal distribution).

7.2 Problem. Create a conceptual map to remark transformers' foundations.

Solution. The transformer's architecture is based on (i) multi-head attention that splits input into multiple heads and computes attention independently. (ii) Scaled Dot-Product Attention that computes attention scores using dot products. (iii) Positional encoding that adds positional information using sine and cosine functions. The transformers include Input Embedding to convert tokens into dense vectors, an Encoder comprising multi-head self-attention, feedforward network, and layer normalization and residual connections, a decoder including masked multi-head self-attention, multi-head encoder-decoder attention, feedforward network, and layer normalization and residual connections, and an output involving a Linear layer with Softmax. If the decoder part is not considered, the transformer is encoder-only. On the contrary, the transformer is decoder-only when the encoder part is not considered. Overall, a transformer operates as follows. ① Tokenize and embed the input. ② Add positional encodings. ③ Pass through the encoder (multi-head self-attention + FFN). ④ Pass through the decoder (masked self-attention + encoder-decoder attention + FFN). ⑤ Generate output probabilities using a linear layer and Softmax. ⑥ Train the model using cross-entropy loss and backpropagation.

7.3 Problem. Classify flows with AE, VAE, and RNN in the UNIV1 dataset available in the book's repository. Compare the models regarding accuracy, balanced accuracy, precision, recall, and F1-score.

Solving process. Follow the methodology defined in Chap. 3.

- Business and data understanding. The goal is to classify flows with AE, VAE, and RNN. The UNIV1 dataset is loaded using the Pandas library.
- Data Engineering.
  - Remove the duplicates and empty instances.
  - Convert to numerical features like IP and MAC addresses.
  - Normalize the UNIV1 dataset with the StandardScaler.
  - Split the normalized data into training and testing sets using 80:20.
- ML model engineering. Build the models using TensorFlow and Keras as follows.
  - Import the TensorFlow and Keras libraries.
  - Build the AE-based model. First, create an encoder with a dense layer and ReLU activation function. Second, develop a decoder using a dense layer and a Sigmoid activation function for the last layer. Third, connect the encoder output to the decoder's first layer.
  - Build the VAE-based model. First, create an encoder with a dense layer and ReLU activation function. Second, use the encoder output to compute the mean and variance. Third, develop a decoder using a dense layer and ReLU activation

function for the hidden layer; the mean and variance are configured in this layer. Fourth, create the decoder's output with a sigmoid activation function.
  - Create the RNN-based model using LSTM, with the last layer using the Sigmoid activation function (useful for binary classification) and the loss function BinaryCrossentropy.

- Quality assurance for ML applications.

  - Evaluate the models using the test set regarding accuracy, balanced accuracy, precision, recall, and F1-score.
  - Print the evaluation results using the Matplotlib library.
  - Discuss the results.

7.4 Problem. Classify flows in the UNIV1 dataset available in the book's repository by using BERT and DistilBERT. Compare the models regarding accuracy, balanced accuracy, precision, recall, and F1-score.

Solving process. Follow the methodology defined in Chap. 3.

- Business and data understanding. The goal is to classify flows with BERT and DistilBERT. The UNIV1 dataset is loaded using the Pandas library.

- Data engineering.

  - Remove the duplicates and empty instances.
  - Categorize features like IP and MAC addresses.
  - Normalize the dataset with the StandardScaler.
  - Split the normalized data into train and test sets using 80:20.

- ML model engineering.

  - The BERT and DistilBERT-based models are built using PyTorch. How the models are pre-trained, it is just necessary to download the models using the transformer libraries.
  - Employ a fully connected layer with a Sigmoid activation function in the last layer and a Binary Cross Entropy loss function. This configuration must be enforced in both models.

- Quality assurance for ML applications.

  - Evaluate the models using the test set regarding accuracy, balanced accuracy, precision, recall, and F1-score.
  - Print the results using the Matplotlib library.
  - Discuss the results.

7.5 Problem. Write FCAPS-centered intents in NLP and translate to policies in the format (If-Then-Do) using prompt engineering.

Solution. Initially, define some intents in NLP. Finally, develop a prompt to translate them to policies in the format If-Then-Do.

- Intents.

  - TypeName: Report_Failure.
  - Description: User reports a network failure.
  - Examples:

    - I have connection issues.
    - The server is not responding.
    - There is a network outage.

  - TypeName: Change_Configuration.
  - Description: User requests a configuration change.
  - Examples:

    - I want to change the IP address.
    - I need to modify the firewall parameters.
    - Update the router configuration.

  - TypeName: Monitor_Performance.
  - Description: User inquires about network performance.
  - Examples:

    - How is the network traffic?
    - I want to check bandwidth usage.
    - Monitor latency and jitter.

  - TypeName: Security_Alert.
  - Description: User reports a security issue.
  - Examples:

    - I have detected suspicious activity.
    - There is a possible DDoS attack.
    - My system has been compromised.

- Prompt. Please, suppose you are an expert in network management based on FCAPS. Given the following attempt at NLP, translate it into a network management policy in the If-Then-Do format.

  - Input.

    - Intent: Report_Failure.
    - Description: The user reports a network failure.
    - Example: I am having connection problems.

  - Output. If the user reports a connection problem. Then, check the network status in the relevant area. Do, run automatic diagnosis and escalate the problem if necessary.

References

1. A. Vaswani, N. Shazeer, N. Parmar, J. Uszkoreit, L. Jones, A. N. Gomez, L. Kaiser, I. Polosukhin, Attention is all you need, in *Advances in Neural Information Processing Systems*, vol. 30 (2017)
2. J. Devlin, M.-W. Chang, K. Lee, K. Toutanova, BERT: pre-training of deep bidirectional transformers for language understanding, in *NAACL-HLT*, MN, USA, June 2–7, in ed. by J. Burstein, C. Doran, T. Solorio (Association for Computational Linguistics, 2019), pp. 4171–4186. https://doi.org/10.18653/V1/N19-1423
3. J. Briskilal, C.N. Subalalitha, An ensemble model for classifying idioms and literal texts using BERT and ROBERTA. Inf. Process. Manage. **59**(1), 1 (2022). ISSN 0306–4573. https://doi.org/10.1016/j.ipm.2021.102756

4. V. Sanh, L. Debut, J. Chaumond, T. Wolf, DistilBERT, a distilled version of BERT: smaller, faster, cheaper and lighter (2019). arXIV:abs/1910.01108, https://api.semanticscholar.org/CorpusID:203626972
5. Z. Lan, M. Chen, S. Goodman, K. Gimpel, P. Sharma, R. Soricut, ALBERT: A lite BERT for self-supervised learning of language representations (2020). https://arxiv.org/abs/1909.11942
6. A. Buscemi, D. Proverbio, ChatGPT versus Gemini vs LLaMA on multilingual sentiment analysis (2024). https://arxiv.org/abs/2402.01715
7. X. Liu, Y. Zheng, Z. Du, M. Ding, Y. Qian, Z. Yang, J. Tang, GPT understands, too. AI Open (2023), pp. 1–12. ISSN 2666–6510. https://doi.org/10.1016/j.aiopen.2023.08.012
8. Y. Zhang, S. Sun, M. Galley, Y.-C. Chen, C. Brockett, X. Gao, J. Gao, J. Liu, B. Dolan, DialoGPT: large-scale generative pre-training for conversational response generation (2020). https://arxiv.org/abs/1911.00536
9. K. Wang, A. Variengien, A. Conmy, B. Shlegeris, J. Steinhardt, Interpretability in the wild: a circuit for indirect object identification in GPT-2 small (2022). arXiv:2211.00593
10. X. Yang, E. Peynetti, V. Meerman, C. Tanner, What GPT knows about who is who (2022). arXiv:2205.07407
11. S. Black, S. Biderman, E. Hallahan, Q. Anthony, L. Gao, L. Golding, H. He, C. Leahy, K. McDonell, J. Phang, et al., GPT-NeoX-20B: an open-source autoregressive language model (2022). arXiv:2204.06745
12. J. Achiam, S. Adler, S. Agarwal, L. Ahmad, I. Akkaya, F.L. Aleman, D. Almeida, J. Altenschmidt, S. Altman, S. Anadkat, et al., *GPT-4 Technical Report* (2023). arXiv:2303.08774
13. C. Raffel, N. Shazeer, A. Roberts, K. Lee, S. Narang, M. Matena, Y. Zhou, W. Li, P.J. Liu, Exploring the limits of transfer learning with a unified text-to-text transformer. J. Mach. Learn. Res. **21**(140), 1–67 (2020)
14. M. Lewis, Y. Liu, N. Goyal, M. Ghazvininejad, A. Mohamed, O. Levy, V. Stoyanov, L. Zettlemoyer, BART: denoising sequence-to-sequence pre-training for natural language generation, translation, and comprehension (2019). arXiv:1910.13461

Part III
Exemplifying Machine Learning in Network Management

The book's third part exemplifies the use of machine learning in network management. Chapter 8 presents a classification model (intrusion detection) performed via supervised learning techniques and improved by explainability techniques.

Improving and Explaining Learning Models 8

8.1 Attack Detection in Internet of Things

This section illustrates how ML models detect security attacks in a public dataset for anomaly detection in MQTT traffic. MQTT is a messaging protocol for IoT. This protocol uses the publish-subscribe communication model optimal for low-bandwidth environments like IoT. According to the methodology discussed in Chap. 2, it is necessary to carry out data understanding, data engineering, machine learning modeling, and evaluation to detect attacks in MQTT traffic. These steps are detailed hereinafter.

8.1.1 Data Understanding and Engineering

Data Understanding
The MQTT dataset comprises 80.893 samples, 66 features, and a target. This dataset is unbalanced, including normal (78.995 samples) and malicious (1.898 samples) traffic containing DoS, Man in the Middle (MITM), and Intrusion Attacks [1]. In DoS, the attacker generates many service requests to occupy excessive service resources to force the server to crash or to make other legal users unable to attain timely service responses [2]. In MITM, the attacker usually intercepts and alters the communication between two parties without their warrant [3]. An intrusion attack involves unauthorized computer systems or network access to compromise confidentiality, integrity, or availability [4].

Data Engineering
Listing 8.1 shows a Python code snippet to realize Data Engineering in the MQTT dataset. (i) The libraries used in the code are imported. (ii) The MQTT dataset is loaded. (iii) Samples (rows) or features (columns) are deleted when all values are NA to reduce data dimensionality. (iv) All columns with less than 5000 rows and the rows with empty data are deleted to constraint filling data. As a result of steps (iii) and (iv), 49 features are discarded and 9.136 instances are removed.hyper@itemfalse (v) All columns related to IP and MAC addresses are deleted; it includes four features. (vi) Labels are assigned for normal (0) and intrusive traffic (1). (vii) The data is normalized with MinMaxScaler; it transforms the values to a specific range without changing the shape of the original distribution. (viii) The traffic classes are counted to determine their balancing. (ix) The dataset is divided into training and test data. (x) As the classes are unbalanced, the training dataset is balanced with SMOTE by generating synthetic samples for the minority class. The test dataset is not balanced to ulteriorly evaluate the models' robustness.

```
#Import Libraries
import matplotlib.pyplot as plt
import numpy as np
import pandas as pd
from sklearn import datasets
import seaborn as sns

#Load the dataset
df = pd.read_csv('/content/Intrusion.csv')
#Delete a row (sample) or column (feature) when all values are NA
df = df.dropna(axis=1, how='all')
```

Supplementary Information The online version contains supplementary material available at https://doi.org/10.1007/978-3-032-15176-6_8.

O. Caicedo Rendon, *Machine Learning for Network Management*, Textbooks in Telecommunication Engineering, https://doi.org/10.1007/978-3-032-15176-6_8

```
#Delete all columns with less than 5000 rows
df = df.dropna(thresh=5000, axis='columns')
#Delete all empty rows
df = df.dropna(axis=0)
#Delete all columns related to IPs and ports
df = df.drop(['ip.src', 'ip.dst', 'eth.src', 'eth.dst'], axis=1)
#Assign label to normal traffic - 0
df.loc[(df.type == 'normal'),'type']='0'
#Assign label to intrusive traffic - 0
df.loc[(df.type == 'intrusion'),'type']='1'

#Normalize data
from sklearn import preprocessing
from sklearn.preprocessing import MinMaxScaler
min_max_scaler = preprocessing.MinMaxScaler()
df_norm = min_max_scaler.fit_transform(df)
df_norm = pd.DataFrame(df_norm)
df_norm = df_norm.rename(columns = {0: 'frame.time_delta', 1: 'frame.time_delta_displayed', 2: 'frame.time_epoch',
                                    3: 'frame.time_relative', 4: 'tcp.srcport', 5: 'tcp.dstport',
                                    6: 'frame.cap_len', 7: 'frame.encap_type', 8: 'frame.ignored',
                                    9: 'frame.len', 10: 'frame.marked', 11: 'frame.number',
                                    12: 'frame.offset_shift', 13: 'type'})

#Count the classes to determine balancing
df.groupby('type').count()

#Divide the dataset
from sklearn.model_selection import train_test_split
from imblearn.over_sampling import SMOTE
from collections import Counter

y=df_norm['type']
X=df_norm.drop(['type'],axis=1)

#Training (80%) and testing (20%) of the dataset
X_train, X_test, y_train, y_test = train_test_split(X, y, test_size=0.20, random_state=42, stratify=y)

#Balancing the training dataset
#Intrusion traffic is increased synthetically until 25% of the normal one
oversample = SMOTE(sampling_strategy=0.25, random_state=42)
over_X_train, over_y_train = oversample.fit_resample(X_train, y_train)

print('Original dataset shape {}'.format(Counter(y_train)))
print('Resampled dataset shape {}'.format(Counter(over_y_train)))
```

Listing 8.1 Attack detection—data engineering

Table 8.1 MQTT dataset, feature description

Field name	Description
frame.time_delta	Delta time since the previously captured frame
frame.time_delta_displayed	Delta time since the previously displayed frame
frame.time_epoch	The frame timestamp from the capture traffic
frame.time_relative	Time elapsed since the first frame
tcp.srcport	Origin TCP port
tcp.dstport	Destination TCP port
frame.cap_len	Length of captured frame
frame.encap_type	Type of encapsulation
frame.ignored	Indicates whether the frame is ignored
frame.len	Frame length
frame.marked	Indicates whether the frame is marked
frame.number	Frame number
frame.offset_shift	Frame shift time
type	Indicates whether a traffic is normal or intrusion

Before SMOTE balancing, the MQTT dataset comprises 71,757 instances (69,863 of class 0 and 1894 of class 1), 13 features, and a target. The MQTT training dataset (80% of the entire dataset) includes 55,890 samples of normal and 1515 of intrusion traffic. After SMOTE balancing, the training MQTT dataset contains 55,890 samples of regular traffic and 13,972 of intrusion traffic. Table 8.1 describes the features used to train and test the ML models built in the following section; the description is based on [5].

8.1.2 Learning Models for Attack Detection

Lists 8.2, 8.3, 8.4, 8.5, 8.6, and 8.7 show Python code snippets to detect attacks with ML models in the MQTT dataset.

Machine Learning Modeling

Listing 8.2 shows a Python code snippet to detect attacks with a Random Forest model in the MQTT dataset. (i) The libraries used in the code are imported. (ii) The Random Forest model is created using the balanced training dataset. (iii) The Random Forest model is evaluated using the accuracy, balanced accuracy (i.e., in binary classification, serves to deal with imbalanced datasets; it is defined as the average of recall obtained on each class), precision, recall, and F1-score metrics. Also, a Confusion Matrix C was computed. C_{ij} is the number of observations known to be in class i and predicted to be in class j. Thus in binary classification, the true negatives are $C_{0,0}$, false negatives are $C_{1,0}$, true positives are $C_{1,1}$, and false positives are $C_{0,1}$.

```
#Import Libraries
from sklearn.metrics import classification_report
from sklearn.ensemble import RandomForestClassifier

#RF Modeling --------------------
rfclf = RandomForestClassifier(n_estimators=100, random_state=42,
    max_depth=5)
rfclf.fit(over_X_train, over_y_train)

#RF Evaluation --------------------
from sklearn.metrics import confusion_matrix, accuracy_score,
    precision_score, recall_score, f1_score
from sklearn.metrics import balanced_accuracy_score
y_predict_rf = rfclf.predict(X_test)
y_predict_rf_train = rfclf.predict(over_X_train)

#Confusion matrix computation and plotting
cnf=confusion_matrix(y_test, y_predict_rf)
class_names=[0,1]
fig, ax = plt.subplots()
tick_marks = np.arange(len(class_names))
plt.xticks(tick_marks, class_names)
```

```
plt.yticks(tick_marks, class_names)

#Heatmap for confusion matrix
sns.heatmap(pd.DataFrame(cnf), annot=True, cmap="YlGnBu" ,fmt='g')
ax.xaxis.set_label_position("top")
plt.tight_layout()
plt.title('Confusion matrix', y=1.1)
plt.ylabel('Actual label')
plt.xlabel('Predicted label')

#Evaluation metrics in testing dataset
print()
print("Test Metrics...")
print()
print("Accuracy :: ", accuracy_score(y_test, y_predict_rf))
print()
print("Balanced Accuracy", balanced_accuracy_score(y_test, y_predict_rf))
print()
print("Precision :: ", precision_score(y_test, y_predict_rf))
print()
print("Recall :: ", recall_score(y_test, y_predict_rf))
print()
print("F1-Score :: ", f1_score(y_test, y_predict_rf))
print()
print(Classification_report(y_test, y_predict_rf))
```

Listing 8.2 Attack detection—random forest

Listing 8.2 shows a Python code snippet to detect attacks with an SVM model in the MQTT dataset. (i) The libraries used in the code are imported. (ii) The SVM model is created using the balanced training dataset and evaluated with accuracy, balanced accuracy, precision, recall, F1-score, and Confusion Matrix.

```
#SVM Modeling with RBF kernel --------------------
from sklearn import svm
svm_model = svm.SVC(kernel='rbf', C=0.5, gamma=0.1, class_weight='balanced')
svm_model.fit(over_X_train, over_y_train)

#SVM Evaluation --------------------
from sklearn.metrics import confusion_matrix, accuracy_score, precision_score, recall_score, f1_score
from sklearn.metrics import balanced_accuracy_score

y_predict_svm = svm_model.predict(X_test)
y_predict_svm_train = svm_model.predict(over_X_train)

#Confusion matrix computation and plotting
cnf=confusion_matrix(y_test, y_predict_svm)
class_names=[0,1]
fig, ax = plt.subplots()
tick_marks = np.arange(len(class_names))
plt.xticks(tick_marks, class_names)
plt.yticks(tick_marks, class_names)

#Heatmap for confusion matrix
sns.heatmap(pd.DataFrame(cnf), annot=True, cmap="YlGnBu" ,fmt='g')
ax.xaxis.set_label_position("top")
plt.tight_layout()
plt.title('Confusion matrix', y=1.1)
plt.ylabel('Actual label')
plt.xlabel('Predicted label')

#Evaluation metrics in testing dataset
print()
print()
print("Test Metrics...")
print()
```

```
print("Accuracy :: ", accuracy_score(y_test, y_predict_svm))
print()
print("Balanced Accuracy", balanced_accuracy_score(y_test, y_predict_svm))
print()
print("Precision :: ", precision_score(y_test, y_predict_svm))
print()
print("Recall :: ", recall_score(y_test, y_predict_svm))
print()
print("F1-Score :: ", f1_score(y_test, y_predict_svm))
print()
print(classification_report(y_test, y_predict_svm))
```

Listing 8.3 Attack detection—SVM

Listing 8.3 shows a Python code snippet to detect attacks with an XGBoost model in the MQTT dataset. (i) The libraries used in the code are imported. (ii) The XGBoost model is created using the balanced training dataset and evaluated with accuracy, balanced accuracy, precision, recall, F1-score, and Confusion Matrix.

```
#XGBoost Modeling --------------------
from xgboost import XGBClassifier
xgbclf = XGBClassifier(learning_rate=0.01, max_depth=5, n_estimators=100)
xgbclf.fit(over_X_train, over_y_train)

#XGBoost Evaluation --------------------
y_predict_xgb = xgbclf.predict(X_test)
y_predict_xgb_train = xgbclf.predict(over_X_train)

#Confusion matrix computation and plotting
cnf=confusion_matrix(y_test, y_predict_xgb)
class_names=[0,1]
fig, ax = plt.subplots()
tick_marks = np.arange(len(class_names))
plt.xticks(tick_marks, class_names)
plt.yticks(tick_marks, class_names)

#Heatmap for confusion matrix
sns.heatmap(pd.DataFrame(cnf), annot=True, cmap="YlGnBu" ,fmt='g')
ax.xaxis.set_label_position("top")
plt.tight_layout()
plt.title('Confusion matrix', y=1.1)
plt.ylabel('Actual label')
plt.xlabel('Predicted label')

#Evaluation metrics in testing dataset
print()
print("Test Metrics...")
print()
print("Accuracy :: ", accuracy_score(y_test, y_predict_xgb))
print()
print("Balanced Accuracy", balanced_accuracy_score(y_test, y_predict_xgb))
print()
print("Precision :: ", precision_score(y_test, y_predict_xgb))
print()
print("Recall :: ", recall_score(y_test, y_predict_xgb))
print()
print("F1-Score :: ", f1_score(y_test, y_predict_xgb))
print()
print(classification_report(y_test, y_predict_xgb))
```

Listing 8.4 Attack detection—XGBoost

Listing 8.4 shows a Python code snippet to detect attacks with a CNN model in the MQTT dataset. (i) The libraries used in the code are imported. (ii) The CNN model is created using the balanced training dataset and evaluated with the accuracy, balanced accuracy, precision, recall, and F1-score metrics. For this model, a checkpoint performs early stops when the accuracy metric does not improve in a n number of epochs.

```
#CNN Modeling --------------------
import keras
import tensorflow as tf
keras.__version__
from keras import backend as K
from keras.models import Sequential
from tensorflow.keras.layers import Embedding, Conv1D, MaxPooling1D, Flatten, Dense, GlobalMaxPooling1D, Dropout
from keras.callbacks import ModelCheckpoint

#CNN structure
#Conv1D is a convolutional layer
#64 filters in the convolutional layer
#Each filter will span 5 steps of the input sequence
#13 Features, 1 feature per step
#Tanh is the activation function for this layer
#Flatten: convolutional layer end - full connected start
#Dropout randomly sets 50% of the inputs to zero at each update during training
#Dropout helps the network generalize better
#3 hidden fully connected layers
#A fully connected output layer with 1 neuron
#Sigmoid is the activation function for this layer

model_cnn = Sequential([
  Conv1D(64, 5, activation='tanh', input_shape=(13, 1)),
  GlobalMaxPooling1D(),
  Flatten(),
  Dropout(0.5),
  Dense(32, activation='tanh'),
  Dense(8, activation='tanh'),
  Dense(2, activation='tanh'),
  Dense(1, activation='sigmoid')
])

#CNN Evaluation --------------------
#These functions allow computing the evaluation metrics
def recall_m(y_true, y_pred):
    true_positives = K.sum(K.round(K.clip(y_true * y_pred, 0, 1)))
    possible_positives = K.sum(K.round(K.clip(y_true, 0, 1)))
    recall = true_positives / (possible_positives + K.epsilon())
    return recall

def precision_m(y_true, y_pred):
    true_positives = K.sum(K.round(K.clip(y_true * y_pred, 0, 1)))
    predicted_positives = K.sum(K.round(K.clip(y_pred, 0, 1)))
    precision = true_positives / (predicted_positives + K.epsilon())
    return precision

def f1_m(y_true, y_pred):
    precision = precision_m(y_true, y_pred)
    recall = recall_m(y_true, y_pred)
    return 2*((precision*recall)/(precision+recall+K.epsilon()))

def balanced_acc_m(y_true, y_pred):
    tp = K.sum(K.round(K.clip(y_true * y_pred, 0, 1)))
    tn = K.sum(K.round(K.clip((1 - y_true) * (1 - y_pred), 0, 1)))
    fp = K.sum(K.round(K.clip((1 - y_true) * y_pred, 0, 1)))
    fn = K.sum(K.round(K.clip(y_true * (1 - y_pred), 0, 1)))

    sensitivity = tp / (tp + fn + K.epsilon())
```

```
    specifity = tn / (fp + tn + K.epsilon())
    return (sensitivity + specifity) / (2 + K.epsilon())

#A checkpoint is defined for determining val_accuracy improvement
#Stop_early when val_accuracy does not improve during 10 epochs

model_cnn.compile(optimizer=tf.keras.optimizers.Adam(learning_rate=0.001)
    ,
                 loss='binary_crossentropy', metrics=['accuracy',
    balanced_acc_m, precision_m, recall_m, f1_m])
checkpoint = ModelCheckpoint("best_epoch_cnn_model.h5", monitor='
    val_accuracy', verbose=1,save_best_only=True, mode='auto', period=1,
    save_weights_only=False)
stop_early = tf.keras.callbacks.EarlyStopping(monitor='val_accuracy',
    patience=10)
history = model_cnn.fit(over_X_train, over_y_train, batch_size=64, epochs
    =50, validation_data=(X_test, y_test), callbacks=[checkpoint,
    stop_early])

#History is a json file that stores metric results per epoch
import json
results_cnn = json.dumps(history.history)
with open('results_cnn_instrusion.json', 'w') as archivo:
    archivo.write(results_cnn)
```

Listing 8.5 Attack detection—CNN

Listing 8.5 shows a Python code snippet to detect attacks with an Autoencoder model in the MQTT dataset. (i) The libraries used in the code are imported. (ii) The Autoencoder model is created using the balanced training dataset and evaluated the accuracy, balanced accuracy, precision, recall, and F1-score metrics. This model uses an early stop as the developed for the CNN model.

```
#Autoencoder Modeling --------------------
from tensorflow.keras import layers
import tensorflow as tf

#Define the input size for the AE
input_dim = X.shape[1]
#12 features (latent layers) - The original dataset includes 13 features
latent_dim = 12

#Create the autoencoder
#Input Layer
input_layer = layers.Input(shape=(input_dim,))
#Encoder Layer
encoded = layers.Dense(64, activation='relu')(input_layer)
encoded = layers.Dense(latent_dim, activation='relu')(encoded)
#Decoder Layer
decoded = layers.Dense(64, activation='relu')(encoded)
decoded = layers.Dense(input_dim, activation='sigmoid')(decoded)
#Autoencoder model
autoencoder = tf.keras.Model(inputs=input_layer, outputs=decoded)
#Encoder model to get the latent representations
encoder = tf.keras.Model(inputs=input_layer, outputs=encoded)
#Compile Autoencoder
autoencoder.compile(optimizer='adam', loss='mse')

#Training the autoencoder
#The over_X_train is provided as both the input and target output
#An autoencoder is trained to reconstruct its input
history = autoencoder.fit(over_X_train, over_X_train,
                          epochs=50,
                          batch_size=64,
                          shuffle=True,
                          validation_data=(X_test, X_test))
```

```
#Get latent representations for training and test datasets
latent_representations_train = encoder.predict(over_X_train)
latent_representations_test = encoder.predict(X_test)

#A NN is used to perform binary classification
classifier_input = layers.Input(shape=(latent_dim,))
classifier_output = layers.Dense(1, activation='sigmoid')(
    classifier_input)

#Autoencoder Evaluation -------------
#These functions allow computing the evaluation metrics
def recall_m(y_true, y_pred):
    true_positives = K.sum(K.round(K.clip(y_true * y_pred, 0, 1)))
    possible_positives = K.sum(K.round(K.clip(y_true, 0, 1)))
    recall = true_positives / (possible_positives + K.epsilon())
    return recall

def precision_m(y_true, y_pred):
    true_positives = K.sum(K.round(K.clip(y_true * y_pred, 0, 1)))
    predicted_positives = K.sum(K.round(K.clip(y_pred, 0, 1)))
    precision = true_positives / (predicted_positives + K.epsilon())
    return precision

def f1_m(y_true, y_pred):
    precision = precision_m(y_true, y_pred)
    recall = recall_m(y_true, y_pred)
    return 2*((precision*recall)/(precision+recall+K.epsilon()))

def balanced_acc_m(y_true, y_pred):
    tp = K.sum(K.round(K.clip(y_true * y_pred, 0, 1)))
    tn = K.sum(K.round(K.clip((1 - y_true) * (1 - y_pred), 0, 1)))
    fp = K.sum(K.round(K.clip((1 - y_true) * y_pred, 0, 1)))
    fn = K.sum(K.round(K.clip(y_true * (1 - y_pred), 0, 1)))

    sensitivity = tp / (tp + fn + K.epsilon())
    specifity = tn / (fp + tn + K.epsilon())
    return (sensitivity + specifity) / (2 + K.epsilon())

#A checkpoint is defined for determining val_accuracy improvement
#Stop_early when val_accuracy does not improve during 10 epochs

from keras.callbacks import ModelCheckpoint
classifier = tf.keras.Model(inputs=classifier_input, outputs=
    classifier_output)
classifier.compile(optimizer=tf.keras.optimizers.Adam(learning_rate
    =0.0001),
                   loss='binary_crossentropy', metrics=['accuracy',
    balanced_acc_m, precision_m, recall_m, f1_m])
checkpoint = ModelCheckpoint("best_epoch_sae_model.h5", monitor='
    val_accuracy', verbose=1, save_best_only=True, mode='auto', period=1,
     save_weights_only=False)
stop_early = tf.keras.callbacks.EarlyStopping(monitor='val_accuracy',
    patience=10)
history = classifier.fit(latent_representations_train, over_y_train,
    batch_size=64, epochs=50, validation_data=(
    latent_representations_test, y_test), callbacks=[checkpoint,
    stop_early])

#Classifier Evaluation
from sklearn.metrics import accuracy_score

y_pred = (classifier.predict(latent_representations_test) > 0.5).astype(
    int)
accuracy = accuracy_score(y_test, y_pred)
print(f"Accuracy del Clasificador: {accuracy:.4f}")
```

```

#History is a json file that stores metric results per epoch
import json
results_sae = json.dumps(history.history)
with open('results_sae_instrusion.json', 'w') as archivo:
    archivo.write(results_sae)
```

Listing 8.6 Attack detection—autoencoder

Listing 8.6 shows a Python code snippet to detect attacks with a RNN model in the MQTT dataset. (i) The libraries used in the code are imported. (ii) The RNN model is created using LSTM and the balanced training dataset and evaluated with the accuracy, balanced accuracy, precision, recall, and F1-score metrics. This model uses an early stop as the CNN model does.

```
#RNN Modeling -----------------------------
#Convert tabular data in sequence data - it is demanded by RNN
def create_sequences(data, target, sequence_length):
    X, y = [], []
    for i in range(len(data) - sequence_length):
        X.append(data[i:i + sequence_length])
        y.append(target[i + sequence_length])
    return np.array(X), np.array(y)

#Sequences length
sequence_length = 10
X_seq, y_seq = create_sequences(X, y, sequence_length)

#Divide the dataset into training and test
X_train, X_test, y_train, y_test = train_test_split(X_seq, y_seq,
    test_size=0.20, random_state=42, stratify=y_seq)

#Data is transformed to the type of input needed by SMOTE
n_samples_train, seq_len, n_features = X_train.shape
X_train_flat = X_train.reshape(n_samples_train, -1)

#Data is balanced with SMOTE
oversample = SMOTE(sampling_strategy=0.25, random_state=42)
over_X_train, over_y_train = oversample.fit_resample(X_train_flat,
    y_train)

#Data is transformed to the type of input needed by RNN
over_X_train = over_X_train.reshape(-1, seq_len, n_features)

print('Original dataset shape {}'.format(Counter(y_train)))
print('Resampled dataset shape {}'.format(Counter(over_y_train)))

#RNN training
from keras.models import Sequential
from keras import layers
from keras import regularizers
from keras import backend as K
from keras.callbacks import ModelCheckpoint
from keras.models import Sequential
from keras.layers import Dense, Embedding, LSTM, Dropout

#RNN structure - LSTM in this example
model_rnn = tf.keras.Sequential([
    tf.keras.layers.LSTM(64, activation='tanh', input_shape=(
    sequence_length, 13)),
    tf.keras.layers.Dense(32, activation='tanh'),
    tf.keras.layers.Dense(8, activation='tanh'),
    tf.keras.layers.Dense(2, activation='tanh'),
    tf.keras.layers.Dense(1, activation='sigmoid')])

#RNN Evaluation -----------------------------
#These functions allow computing the evaluation metrics
def recall_m(y_true, y_pred):
```

```
    true_positives = K.sum(K.round(K.clip(y_true * y_pred, 0, 1)))
    possible_positives = K.sum(K.round(K.clip(y_true, 0, 1)))
    recall = true_positives / (possible_positives + K.epsilon())
    return recall

def precision_m(y_true, y_pred):
    true_positives = K.sum(K.round(K.clip(y_true * y_pred, 0, 1)))
    predicted_positives = K.sum(K.round(K.clip(y_pred, 0, 1)))
    precision = true_positives / (predicted_positives + K.epsilon())
    return precision

def f1_m(y_true, y_pred):
    precision = precision_m(y_true, y_pred)
    recall = recall_m(y_true, y_pred)
    return 2*((precision*recall)/(precision+recall+K.epsilon()))

def balanced_acc_m(y_true, y_pred):
    tp = K.sum(K.round(K.clip(y_true * y_pred, 0, 1)))
    tn = K.sum(K.round(K.clip((1 - y_true) * (1 - y_pred), 0, 1)))
    fp = K.sum(K.round(K.clip((1 - y_true) * y_pred, 0, 1)))
    fn = K.sum(K.round(K.clip(y_true * (1 - y_pred), 0, 1)))

    sensitivity = tp / (tp + fn + K.epsilon())
    specifity = tn / (fp + tn + K.epsilon())
    return (sensitivity + specifity) / (2 + K.epsilon())

#A checkpoint is defined for determining val_accuracy improvement
#Stop_early when val_accuracy does not improve during 10 epochs

model_rnn.compile(optimizer=tf.keras.optimizers.Adam(learning_rate=0.001),
                  loss='binary_crossentropy', metrics=['accuracy', balanced_acc_m, precision_m, recall_m, f1_m])
checkpoint = ModelCheckpoint("best_epoch_rnn_model.h5", monitor='val_accuracy', verbose=1, save_best_only=True, mode='auto', period=1, save_weights_only=False)
stop_early = tf.keras.callbacks.EarlyStopping(monitor='val_accuracy', patience=10)
history = model_rnn.fit(over_X_train, over_y_train, batch_size=64, epochs=50, validation_data=(X_test, y_test), callbacks=[checkpoint, stop_early])

#History is a json file that stores metric results per epoch
import json
results_rnn = json.dumps(history.history)
with open('results_rnn_intrusion.json', 'w') as archivo:
    archivo.write(results_rnn)
```

Listing 8.7 Attack detection—RNN

8.1.3 Learning Models' Evaluation

Results and Analysis

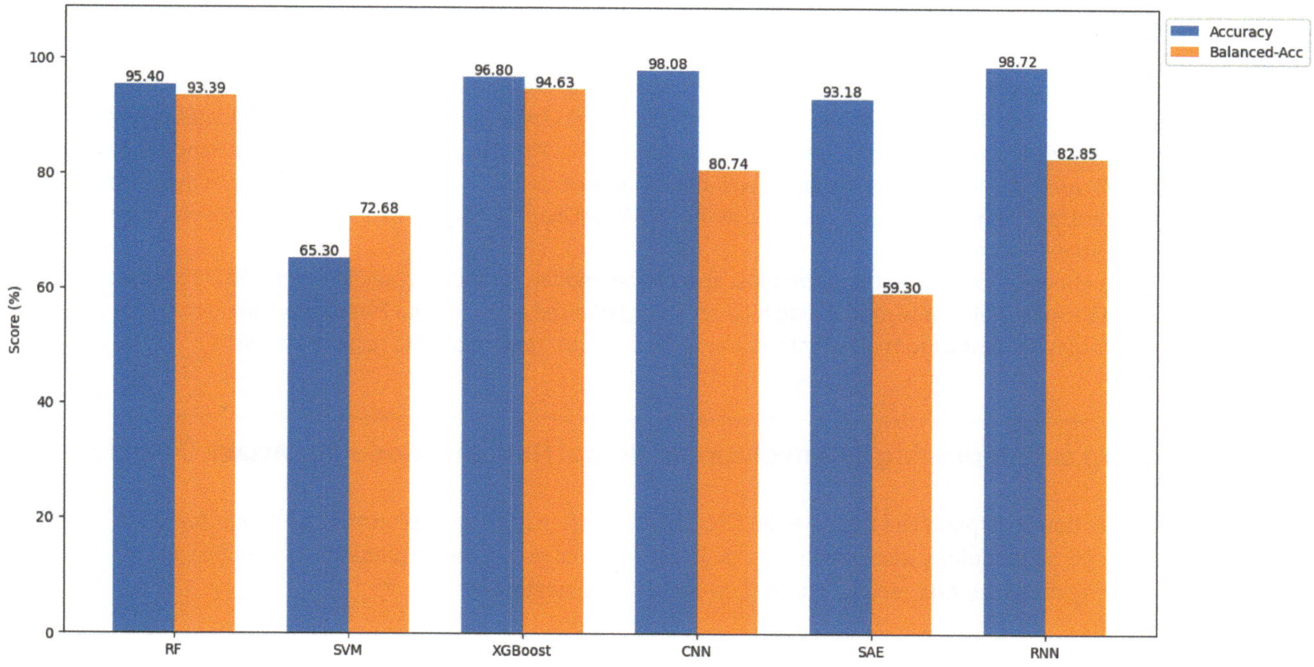

Fig. 8.1 Models comparison regarding accuracy and balanced accuracy

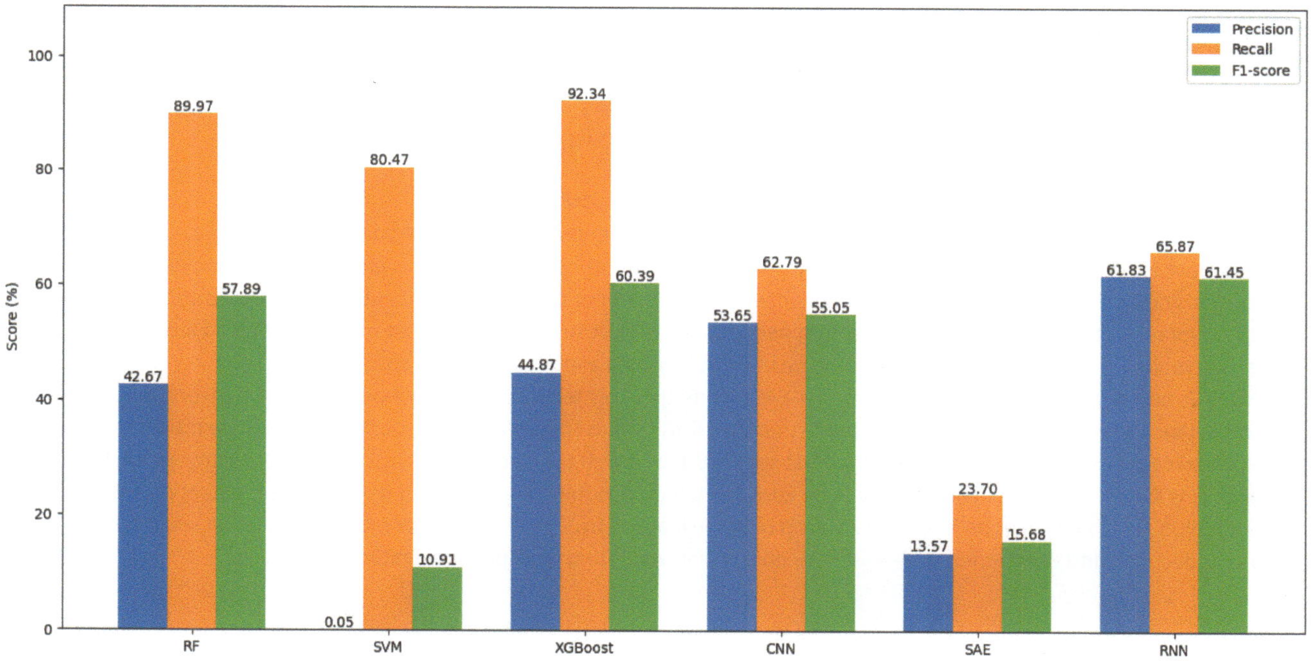

Fig. 8.2 Models comparison regarding precision, recall, and F1-score

Figure 8.1 shows the results of Random Forest, SVM, XGBoost, Autoencoder, CNN, and RNN models regarding accuracy and balanced accuracy. The RNN and CNN models obtained higher accuracy (98.72 and 98.08%) than the other evaluated models and a good balanced accuracy (82.85 and 80.74%). These results show that the RNN and CNN models detected regular traffic better than intrusion traffic. The RF and XGBoost models got a good trade-off between accuracy (95.40 and 93.39%) and balanced accuracy (96.80 and 94.63%), revealing that these models behave suitably with unbalanced data as that included in the MQTT dataset. The AE model obtained high accuracy (93.18%) and a balanced accuracy low (59.30%). The SVM model got accuracy (65.30%) and balanced accuracy (72.68%) lows. According to the results above, the RNN,

XGBoost, Random Forest, and CNN models are, in order, better candidates than SVM and AE for detecting attacks in the unbalanced MQTT test dataset. Remarkably, RNN and CNN models should be improved in terms of balanced accuracy. Furthermore, performing an overfitting analysis for XGBoost and Random Forest models would be necessary.

Figure 8.2 shows the results of Random Forest, XGBoost, SVM, Autoencoder, CNN, and RNN models regarding precision (i.e., the quality of predicting a true positive), recall (i.e., the capacity of the model to identify all data points in a relevant class), and F1-score (i.e., combining recall and precision using the harmonic mean). The Autoencoder and SVM models achieved lower F1-scores (15.68 and 10.91%) than the other evaluated ones, corroborating that they are not appropriate candidates for attack detection in the MQTT dataset. The RNN and XGBoost models are the best candidates regarding the F1-score metric since they obtained higher scores (61.45 and 60.39%) than the CNN (55.05%) and Random Forest (57.89%) models, meaning that they can identify well class 0 (normal traffic) while minimizing false positives (class 0 identified as 1) and false negatives (class 1 identified as 0). The RNN, XGBoost, CNN, and Random Forest models should be improved to achieve F1-scores >70.

In summary, the above results suggest that the most appropriate models for attack detection in the MQTT dataset are the RNN and XGBoost models. These models had high accuracy, balanced accuracy, and F1-score, meaning that they can classify well both classes: class 0 (normal traffic) and class 1 (anomalous traffic related to security attacks).

8.2 Explaining and Improving a Convolutional Neural Network Model for Attack Detection

This section illustrates how to explain and improve the CNN model built in the previous section for attack detection in MQTT traffic. According to the methodology discussed in Chap. 2, carrying out the Given Explanation and Improve Model steps is necessary after building the black-box model. These steps are detailed hereinafter.

8.2.1 Enforcing Explainability

Data Understanding, Data Engineering, and Machine Learning Modeling

The construction of the black-box model is explained in detail in Sects. 8.1.2 and 8.1.1. In particular, Listing 8.1 shows the Python code snippet that realizes data engineering, and Listing 8.5 presents the Python code snippet that builds, trains, and evaluates the CNN model for attack detection in the MQTT dataset.

SHapley Additive exPlanations for Convolutional Neural Networks

Overall, three categories of XAI techniques can be enforced depending on the ML model phase: in-model, pre-model, and post-model. The in-model is carried out by using transparent models. Pre-model is enforced before model training, comprising explainability algorithms that enhance understanding of the dataset used in the models' construction, such as Exploratory Data Analysis (EDA), feature engineering, and dataset summarization. Post-model is enforced after training an ML model, including techniques that provide insights into how a model generates specific results.

SHAP is one of the most popular post-model XAI techniques. This technique uses cooperative game theory to assign importance to each feature in a model's prediction via a SHAP value. The SHAP technique assumes that the model can be approximated as a sum of feature contributions; see Eq. 8.1, where $f(x)$ is the model output, ϕ_0 is the average prediction, and ϕ_i is the SHAP value of feature i given by Eq. 8.2 in which F is the complete set of features, S is a subset of features excluding i, $f(S)$ is the model's prediction using only the features in S, $f(S \cup \{i\})$ is the model's prediction when feature i is added to S, and the fraction $\frac{|S|!(|F|-|S|-1)!}{|F|!}$ ensures fair weighting of all possible feature subsets.

$$f(x) = \phi_0 + \sum_{i=1}^{n} \phi_i \tag{8.1}$$

$$\phi_i = \sum_{S \subseteq F \setminus \{i\}} \frac{|S|!(|F| - |S| - 1)!}{|F|!} [f(S \cup \{i\}) - f(S)] \tag{8.2}$$

```
#Import libraries
import tensorflow as tf
import pickle
import shap

#Load the CNN model built
metric_list = {"balanced_acc_m": balanced_acc_m, "precision_m": precision_m, "recall_m": recall_m, "f1_m": f1_m}
load_cnn_model = tf.keras.models.load_model('/content/best_epoch_cnn_model.h5', custom_objects=metric_list)

#Test the loaded model
y_predict_cnn = load_cnn_model.evaluate(X_test, y_test, verbose=0)
y_predict_cnn

#SHAP values are computed with 10% of training dataset
distrib_samples_train = shap.utils.sample(over_X_train, 5589, random_state=42)
distrib_samples_test = shap.utils.sample(X_test, 1397, random_state=42)

#SHAP values are stored (to reuse) and loaded
filename_expl = 'shap_values_cnn.sav'
pickle.dump(shap_values, open(filename_expl, 'wb'))
filename_expl_load = '/content/shap_values_cnn.sav'
shap_values = pickle.load(open(filename_expl_load, 'rb'), )
print(shap_values)
feature_names = ['frame.time_delta', 'frame.time_delta_displayed','frame.time_epoch',
                 'frame.time_relative', 'tcp.srcport', 'tcp.dstport',
                 'frame.cap_len', 'frame.encap_type', 'frame.ignored',
                 'frame.len', 'frame.marked', 'frame.number',
                 'frame.offset_shift']

#SHAP results are plotted
shap.summary_plot(shap_values=shap_values[0], features=distrib_samples_test, feature_names=feature_names)
```

Listing 8.8 Enforcing SHAP in CNN

Listing 8.8 shows a Python code snippet to explain by SHAP the CNN model built for attack detection in the MQTT dataset. (i) The libraries used in the code are imported. (ii) The CNN model previously saved (best_epoch_cnn_model.h5, see Listing 8.5) is loaded and tested. (iii) The SHAP values are computed; in the example, just for resource scarcity, the 10% of training MQTT dataset is used. (iv) SHAP results are stored and plotted.

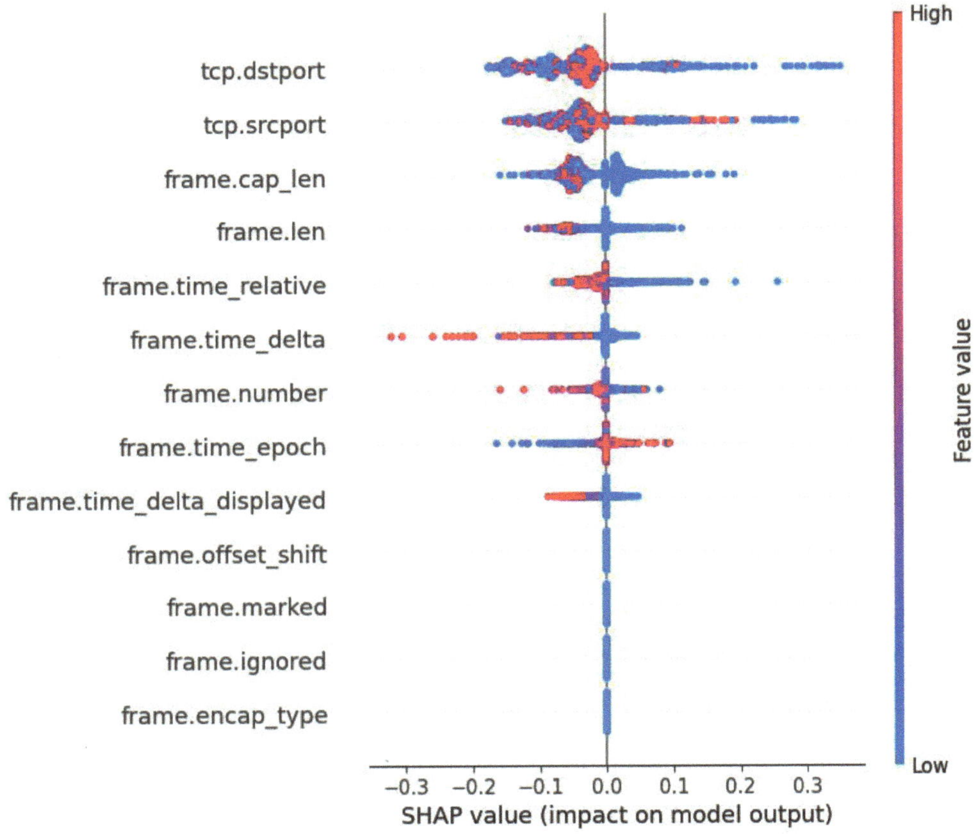

Fig. 8.3 SHAP summary

Figure 8.3 highlights the most relevant features for the CNN to determine whether the traffic is normal or intrusive. The most important one is tcp.dstport, where higher values (a more intense red color) are more associated with class 0 traffic. For the feature frame.time_relative, it is observed that higher values are associated with intrusive traffic. For the feature frame.time_delta_displayed, high red values are related to class 0, while low values are associated with class 1 traffic. The four least relevant features for the CNN model are frame.encap_type, frame.ignored, frame.marked, and frame.offset_shift since their SHAP values are zero.

8.2.2 Improving the Black-Box Model

Enhancing the CNN for Attack Detection

The SHAP results (see Fig. 8.3 suggest removing the features: frame.encap_type, frame.ignored, frame.marked, and frame.offset_shift. After such a removal, see Figs. 8.4 and 8.5, the CNN model improves negligibly in accuracy (from 98.08 to 98.49) and considerably in balanced accuracy (from 80.74 to 88.07) and F1-score (from 55.05 to 65.01). With this improvement, CNN becomes a feasible option for attack detection in the MQTT dataset; their results become equal to or better than the offered by RNN and XGBoost models; it is needed to clarify that those models can also be improved via XAI techniques like SHAP and LIME.

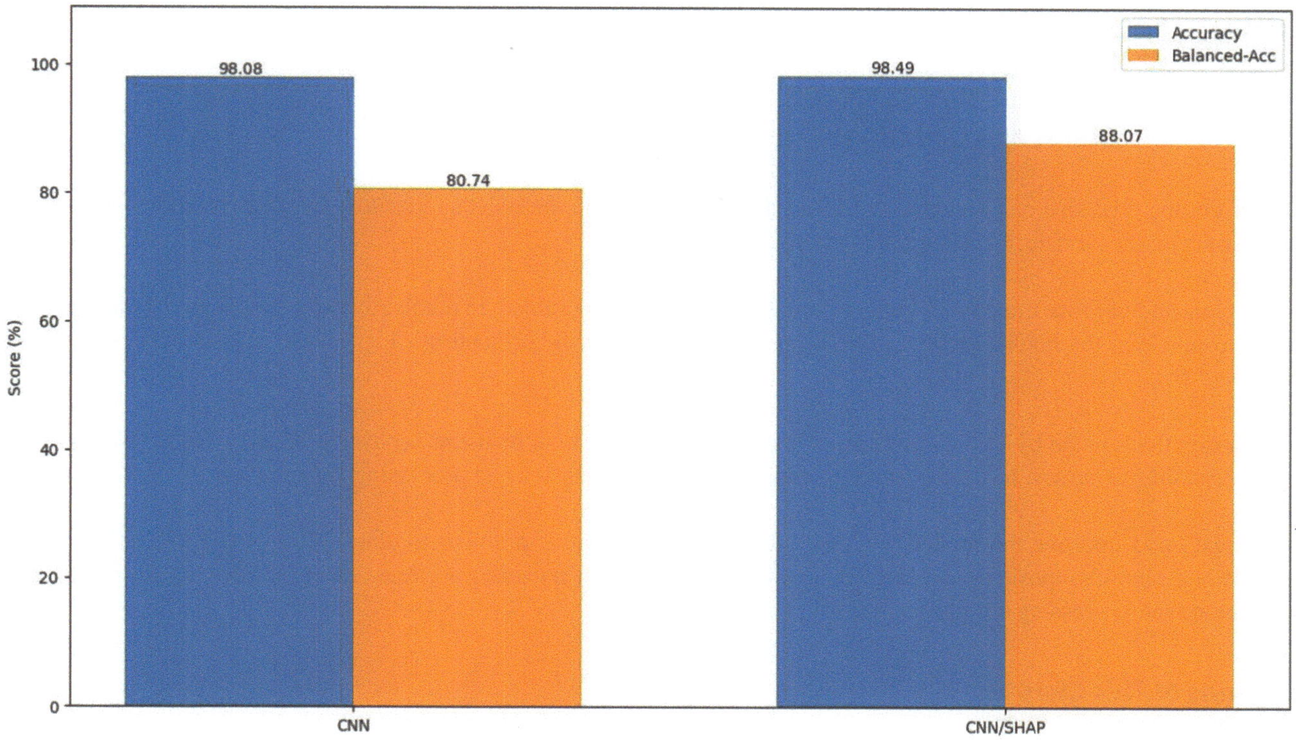

Fig. 8.4 CNN improved: accuracy and balanced accuracy

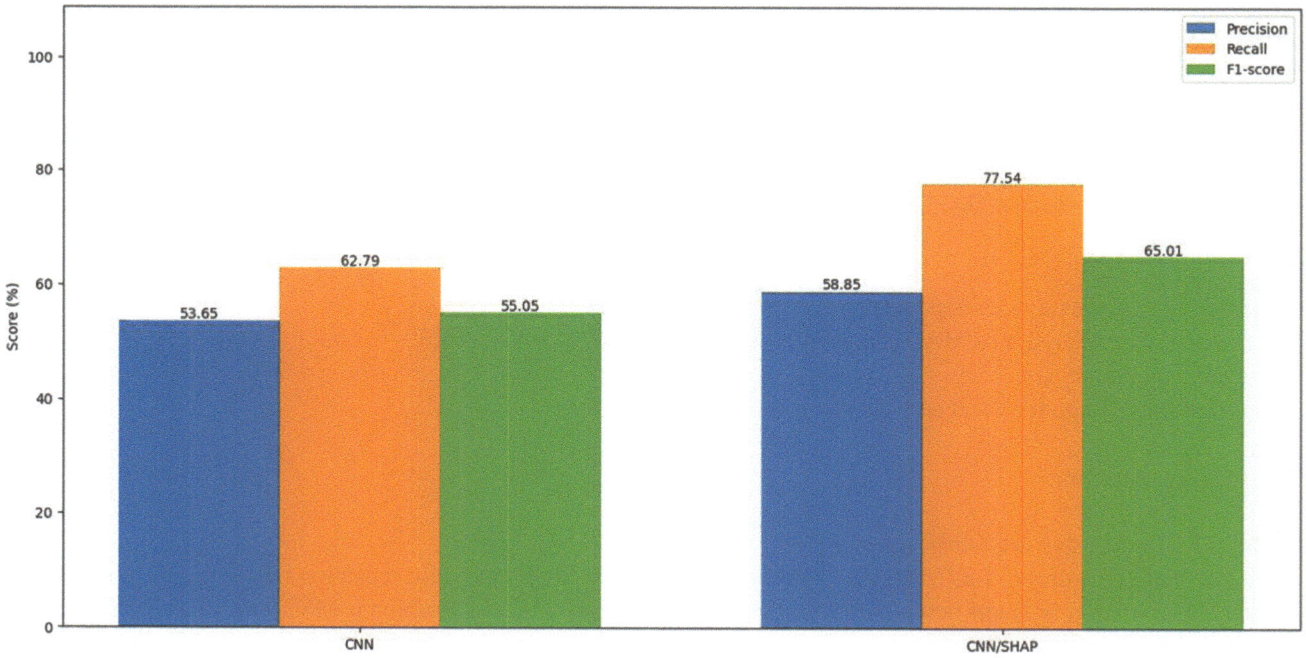

Fig. 8.5 CNN improved: precision, recall, and F1-score

8.3 Summary

This chapter, initially, exemplified the construction of different ML models for attack detection in the MQTT dataset. Then, it enforced a XAI technique (i.e., SHAP) in one of the black-box models built (i.e., the CNN model) to perform feature relevance. Finally, it enhanced the CNN model considering the results from the SHAP analysis.

Homework Problems and Questions

8.1 Predict the total size of packets in the UNI1 dataset available in the book's repository by using XGBoost, CNN, and RNN. Compare the models regarding MSE, RMSE, and R^2.

8.2 Explain and improve the models built as solution in the previous problem to predict the total size of packets in the UNI1 dataset available in the book's repository by using SHAP. Compare the original models and the improved ones regarding RMSE and R^2.

8.3 Classify flows in the UNIV1 dataset available in the book's repository by using Zephyr-7b-beta and BERT. Compare the models regarding accuracy, balanced accuracy, response time, and memory and processing consumption.

8.4 Explain and improve the models built as solution in the previous problem to classify flows in the UNIV1 dataset available in the book's repository by using SHAP. Compare the models regarding accuracy, balanced accuracy, response time, and memory and processing consumption.

Solutions of This Chapter

8.1 Problem. Predict the total size of packets in the UNI1 dataset available in the book's repository by using XGBoost, CNN, and RNN. Compare the models regarding MSE, RMSE, and R^2.

Solving Process. For facing this problem, overall it is necessary to follow the four first phases of methodology discussed in Chap. 2, i.e., business and data understanding, data engineering, machine learning model engineering, and quality assurance for machine learning applications.

- Import the Tensorflow and Keras libraries to build the regression models based on XGBoost, CNN, and RNN.
- Business and data understanding. Goal is to predict the flows size. The UNI1 dataset is loaded using the Pandas library.
- Data engineering.
  - Remove the duplicates and empty instances.
  - Convert features like IP and MAC addresses to numerical values.
  - Set the target to tot_size.
  - Split the dataset into train and test sets using 80:20.
- ML model engineering.
  - Build XGBoost using the XGBRegressor.
  - Create the RNN with two LSTM layers, followed by a Dense layer with the Relu activation function. The last layer is a simple Dense layer with one neuron (essential in regressions task). The loss function is MSE with Adam optimizer.
  - Build the CNN with two Conv1D layers, followed by a Dense layer with the Relu activation function. The last layer is a simple Dense layer with one neuron (essential in regressions task). The loss function is MSE with Adam optimizer.
- Quality assurance for ML applications.
  - Evaluate the models using the test set regarding MSE, RMSE, and R^2.
  - Print the evaluation results using the Matplotlib library to ulterior analysis.

8.2 Problem. Explain and improve the models built as solution in the previous problem to predict the total size of packets in the UNI1 dataset available in the book's repository by using SHAP. Compare the original models and the improved ones regarding RMSE and R^2.

Solving Process. For facing this problem, overall it is necessary to get explanations about the models, improve them, and compare the original models and the improved ones regarding MSE, RMSE, and R^2.

- Import the SHAP library to explain the models.
- Given Explanations.
  - Explain the XGBRegressor-based model, built as solution to predict flows size, using the SHAP Explainer that takes the model already trained to compute the Shapley values. Remarkably, the test and train data must be converted to a numpy array since this format is compatible with SHAP Explainer.
  - Visualize the explanations using the SHAP summary plot.
- Improvement.
  - Select and remove the less relevant features from UNI1 by considering the information provided by SHAP summary.
  - Retrain the XGBRegressor-based model using the new UNI1 dataset to improve its performance.
  - Evaluate the model regarding MSE, RMSE, and R^2 metrics obtained after and before SHAP enforcement.
  - Print the evaluation results to determine how much the XGBRegressor-based model improved.
- Given Explanations.
  - Explain the RNN-based model, built as solution to predict flows, size using the SHAP Explainer that takes the model already trained to compute the Shapley values. Remarkably, the test and train data must be converted to a numpy array since this format is compatible with SHAP Explainer.
  - Visualize the explanations using the SHAP summary plot.
- Improvement.
  - Select and remove the less relevant features from UNI1 by considering the information provided by SHAP summary.
  - Retrain the RNN-based model using the new UNI1 dataset to improve its performance.
  - Evaluate the model regarding MSE, RMSE, and R^2 metrics obtained after and before SHAP enforcement.
  - Print the evaluation results to determine how much the RNN-based model improved.
- Given Explanations.
  - Explain the CNN-based model, built as solution to predict flows size, using the SHAP Explainer that takes the model already trained to compute the Shapley values. Remarkably, the test and train data must be converted to a numpy array since this format is compatible with SHAP Explainer.
  - Visualize the explanations using the SHAP summary plot.
- Improvement
  - Select and remove the less relevant features from UNI1 by considering the information provided by SHAP summary.
  - Retrain the CNN-based model using the new UNI1 dataset to improve its performance.
  - Evaluate the model regarding MSE, RMSE, and R^2 metrics obtained after and before SHAP enforcement.
  - Print the evaluation results to determine how much the CNN-based model improved.

8.3 Problem. Classify flows in the UNIV1 dataset available in the book's repository by using Zephyr-7b-beta and BERT. Compare the models regarding accuracy, balanced accuracy, response time, and memory and processing consumption.

Solving Process. For facing this problem, overall it is necessary to follow the four first phases of methodology discussed in Chap. 2, i.e., business and data understanding, data engineering, machine learning model engineering, and quality assurance for machine learning applications.

- Import the Torch and Transformers libraries to build the models.
- Follow the methodology discussed in Chap. 2.
  - Business and data understanding. Goal defined, classify flows. Data defined, the UNIV1 dataset is loaded using the Pandas library.
  - Data engineering.
    - Remove the duplicates and empty instances.
    - Convert features like IP and MAC addresses to numerical values.
    - Normalize the dataset with StandardScaler.
    - Split the normalized data into train and test sets using 80:20.
  - ML model engineering.
    - Build the Zephyr-7b-beta-based model using PyTorch. How the model is pre-trained, it is just necessary to download it using the transformer library. Employ a fully connected layer with a Sigmoid activation function in the last layer and a Binary Cross Entropy loss function. Enforce the before-described configuration in the model.
    - Build the BERT-based model using PyTorch. How the model is pre-trained, it is just necessary to download it using the transformer library. Employ a fully connected layer with a Sigmoid activation function in the last layer and a Binary Cross Entropy loss function. Enforce the before-described configuration in the model.
  - Quality assurance for ML applications.
    - Evaluate the Zephyr-7b-beta-based and BERT-based models using the test set regarding accuracy, balanced accuracy, response time, and memory and processing consumption.
    - Print the evaluation results using the Matplotlib library to ulterior analysis.

8.4 Problem. Explain and improve the models built as solution in the previous problem to classify flows in the UNIV1 dataset available in the book's repository by using SHAP. Compare the models regarding accuracy, balanced accuracy, response time, and memory and processing consumption.

Solving Process. For facing this problem, overall it is necessary to get explanations about the models, improve them, and compare the original models and the improved ones regarding accuracy, balanced accuracy, response time, and memory and processing consumption.

- Import the SHAP library to explain the models.
- Given explanations.
  - Explain the BERT-based model, built as solution to classify flows, using the SHAP Explainer that takes the model already trained to compute the Shapley values. Remarkably, the test and train data must be converted to a numpy array since this format is compatible with SHAP Explainer.
  - Visualize the explanations using the SHAP summary plot.

- Improvement.
  - Select and remove the less relevant features from UNIV1 by considering the information provided by SHAP summary.
  - Retrain the BERT-based model using the new UNIV1 dataset to improve its performance.
  - Evaluate the model regarding accuracy, balanced accuracy, response time, and memory and processing obtained after and before SHAP enforcement.
  - Print the evaluation results to determine how much the BERT-based model improved.
- Given Explanations.
  - Explain the Zephyr-7b-beta-based model, built as solution to classify flows, using the SHAP Explainer that takes the model already trained to compute the Shapley values. Remarkably, the test and train data must be converted to a numpy array since this format is compatible with SHAP Explainer.
  - Visualize the explanations using the SHAP summary plot.
- Improvement
  - Select and remove the less relevant features from UNIV1 by considering the information provided by SHAP summary.
  - Retrain the Zephyr-7b-beta-based model using the new UNIV1 dataset to improve its performance.
  - Evaluate the model regarding accuracy, balanced accuracy, response time, and memory and processing obtained after and before SHAP enforcement.
  - Print the evaluation results to determine how much the Zephyr-7b-beta-based model improved.

References

1. J.F. Chaves-Tibaduiza, A.I. Becerra-Muoz, A. Robledo-Giron, O.M. Caicedo, On the feasibility of using an encoder-only model for anomaly detection: the bertad approach, in *Proceedings of the 2024 IEEE Colombian Conference on Communications and Computing (COLCOM)* (2024), pp. 1–6. https://doi.org/10.1109/COLCOM62950.2024.10720312
2. I. Ioannou, P. Nagaradjane, P. Angin, P. Balasubramanian, K.J. Kavitha, P. Murugan, V. Vassiliou, Gemlids-miot: a green effective machine learning intrusion detection system based on federated learning for medical iot network security hardening. Comput. Commun. **218**, 209–239 (2024). https://doi.org/10.1016/j.comcom.2024.02.023
3. M.L. Hernandez-Jaimes, A. Martinez-Cruz, K.A. Ramrez-Gutirrez, A machine learning approach for anomaly detection on the internet of things based on locality-sensitive hashing. Integration **96**, 2159 (2024). https://doi.org/10.1016/j.vlsi.2024.102159
4. A. Michelena, M.T.G. Ordas, J. Aveleira-Mata, D.Y.M.D. Blanco, M.T. Diaz, F. Zayas-Gato, E. Jove, J.-L. Casteleiro-Roca, H. Quintian, H. Alaiz-Moreton, J.L. Calvo-Rolle, Beta hebbian learning for intrusion detection in networks with MQTT protocols for IoT devices. Logic J. IGPL **32**, 352–365 (2024)
5. D. Narciandi-Rodríguez, J. Aveleira-Mata, A.M. Corcoba, M. Rubiños, P. Arcano-Bea, H. Alaiz-Moretón, Herramienta de disección de tramas para protocolos iot. Jornadas de Automática **45**, 152 (2024)

Index

O. Caicedo Rendon, *Machine Learning for Network Management*, Textbooks in Telecommunication Engineering, https://doi.org/10.1007/978-3-032-15176-6

FSC
www.fsc.org
MIX
Papier | Fördert gute Waldnutzung
FSC® C083411

Kolibri 360 GmbH
Ferdinand-Jühlke-Straße 7
99095 Erfurt, Deutschland
produktsicherheit@kolibri360.de